Springer Tracts in Mechanical Engineering

Springer Tracts in Mechanical Engineering (STME) publishes the latest developments in Mechanical Engineering - quickly, informally and with high quality. The intent is to cover all the main branches of mechanical engineering, both theoretical and applied, including:

- Engineering Design
- Machinery and Machine Elements
- Mechanical structures and Stress Analysis
- Automotive Engineering
- Engine Technology
- Aerospace Technology and Astronautics
- Nanotechnology and Microengineering
- Control, Robotics, Mechatronics
- MEMS
- Theoretical and Applied Mechanics
- Dynamical Systems, Control
- Fluids mechanics
- Engineering Thermodynamics, Heat and Mass Transfer
- Manufacturing
- Precision engineering, Instrumentation, Measurement
- Materials Engineering
- Tribology and surface technology

Within the scopes of the series are monographs, professional books or graduate textbooks, edited volumes as well as outstanding Ph.D. theses and books purposely devoted to support education in mechanical engineering at graduate and post-graduate levels.

Indexed by SCOPUS and Springerlink. The books of the series are submitted for indexing to Web of Science.

To submit a proposal or request further information, please contact: Dr. Leontina Di Cecco Leontina.dicecco@springer.com or Li Shen Li.shen@springer.com.

Please check our Lecture Notes in Mechanical Engineering at http://www.springer.com/series/11236 if you are interested in conference proceedings. To submit a proposal, please contact Leontina.dicecco@springer.com and Li.shen@springer.com.

More information about this series at http://www.springer.com/series/11693

Ping Jiang · Qi Zhou · Xinyu Shao

Surrogate Model-Based Engineering Design and Optimization

Springer

Ping Jiang
The State Key Laboratory of Digital Manufacturing Equipment and Technology
School of Mechanical Science and Engineering
Huazhong University of Science and Technology
Wuhan, China

Qi Zhou
School of Aerospace Engineering
Huazhong University of Science and Technology
Wuhan, Hubei, China

Xinyu Shao
The State Key Laboratory of Digital Manufacturing Equipment and Technology
School of Mechanical Science and Engineering
Huazhong University of Science and Technology
Wuhan, China

ISSN 2195-9862 ISSN 2195-9870 (electronic)
Springer Tracts in Mechanical Engineering
ISBN 978-981-15-0730-4 ISBN 978-981-15-0731-1 (eBook)
https://doi.org/10.1007/978-981-15-0731-1

This Springer imprint is published by the registered company Springer Nature Singapore Pte Ltd.
The registered company address is: 152 Beach Road, #21-01/04 Gateway East, Singapore 189721, Singapore

Preface

Numerical simulations have been wildly used in engineering design and optimization to study or imitate the real physical systems. With the increasing computational power and the development of simulation software, simulation models can reflect more details of the real system and the discrepancy between the physical system and the simulation model is shrinking. However, conducting a high-fidelity simulation is still time-consuming due to the increasing complexity of the simulation model, for example, the utilization of more solving equations and finer mesh, etc. If directly applying these time-consuming simulation models in optimization problems may also result in unaffordable design cost. Under this background, surrogate models have gained increasing attention in recent years. They are constructed based on available input parameter values and the corresponding output performance or quantity of interests (QOIs) to provide predictions. Especially the multi-fidelity surrogate models, which integrate the information from the simulation model of different fidelity, can achieve a high modeling accuracy within the same simulation cost. Applying surrogate models to engineering design and optimization can significantly reduce the required number of simulations and shorten the design cycle.

For this book, our motivation is to provide a systematic introduction for the designers about how to use surrogate models in engineering design and optimization. It mainly consists of two parts: how to construct a surrogate model with the desired accuracy and how to apply it to different optimization problems. Therefore, this book will cover some of the most popular methods in design space sampling, the ensemble of the surrogate model, multi-fidelity surrogate model construction, surrogate model selection, and validation, surrogate-based robust design optimization, and surrogate model-based evolutionary optimization. Some real-life engineering design problems, such as three-dimensional aircraft design, are also provided to illustrate the ability of Surrogate models in support of complex engineering design. Also, lots of illustrative examples are adopted throughout the book in the hope that the approaches are explained more clearly in this way. We believe that the methods for the ensemble of the surrogate model and multi-fidelity model will be particular interest to the readers, as the presented methods demonstrate a good balance between surrogate modeling accuracy and building cost. This book introduces different approaches in a didactic way, and it can be used as a reference book for undergraduate students, graduate students, or engineers. We assume the intended readers have some mathematical backgrounds, for example, matrix operations, basic optimization algorithms and so on.

We thank Prof. Qi Zhou for his laborious work, Prof. Xinyu Shao for his valuable comments, and the following graduate students: Jiexiang Hu, Leshi Shu, Tingli Xie, Hua Wei, Xiongfeng Ruan, Ji Cheng, Yutong Peng, Linjun Zhong, and Yuda Wu. Ping Jiang, the first author, acknowledges financial support from the National Natural Science Foundation of China (NSFC) under Grant No. 51775203. Qi Zhou, the second author, is grateful for the financial support from the National Natural Science Foundation of China (NSFC) under Grant No. 51805179. Xinyu Shao, the third author, is grateful for the financial support from the National Natural Science Foundation of China (NSFC) under Grant No. 51721092.

Wuhan, China
August 2019

Ping Jiang

Contents

Chapter 1
Introduction

1.1 Reasons for Using Surrogate Models

Simulation models have been widely used to study and analyse complex real-world systems in the design of many modern products, such as vehicles, civil structures and medical devices (Zhou et al. 2016a, 2017). Various types of engineering tasks utilize simulation models, including design space exploration, design optimization, performance prediction, operational management, sensitivity analysis and uncertainty analysis. There are also various problems, such as model calibration and model parameter sensitivity analysis, related to enhancing the ability of simulation models to faithfully reproduce real-world systems (Razavi et al. 2012).

Simulation models can help to predict the performance of systems to facilitate exploration of the design space and search for an optimal design. These simulation models, often implemented with computer codes (e.g. computational fluid dynamics and finite element analysis), can be computationally expensive (Zheng et al. 2013). While the capacity of computers continues to increase, to capture real-world systems more accurately, current simulation codes are becoming even much more complex and unavoidably more expensive (Zheng et al. 2013). It is still impractical to rely only on high-fidelity simulations to yield the full-scale relationships between the design variables as the inputs and the system performance as the output (Christelis and Mantoglou 2016; Zhou et al. 2016b). For example, it can take days to simulate the collapse behaviour of a complete ship structure using finite element analysis or, for the mechanical behaviour of polycrystalline alloy specimens for strength testing using full atomic molecular dynamics simulations, to produce one single output of performance prediction on a computer cluster. Design optimization requires such simulations to be iteratively run and performance predictions to be made for various combinations of input values. Taking Ford Motor Company as an example, it has been reported that it takes the company approximately 36–160 h to run one crash simulation for a full passenger car. For a two-variable optimization problem, under the assumptions that 50 iterations on average are needed for

P. Jiang et al., *Surrogate Model-Based Engineering Design and Optimization*,
Springer Tracts in Mechanical Engineering,
https://doi.org/10.1007/978-981-15-0731-1_1

optimization and that each iteration requires one crash simulation, the total computation time would be 75 days to 11 months (Crombecq et al. 2011). Therefore, relying on full-scale high-fidelity simulations to search for optimal designs is computationally prohibitive.

An effective way to reduce the search time is to utilize surrogate models, also known as approximation models or surrogate models. Such a model acts as a model of a model and thus can replace an expensive simulation model by approximating its input–output responses (Kleijnen 1987; Viana et al. 2014; Tyan et al. 2015). There are many commonly used surrogate modelling techniques (Van Gelder et al. 2014), such as polynomial response surface (PRS) models (Kleijnen 2008), kriging models (Kleijnen 2009; Xiao et al. 2012), neural network models (Can and Heavey 2012) and radial basis function (RBF) models (Fang and Horstemeyer 2006). Details of these types of surrogate models are introduced in Chap. 2.

It is important to note that the quality of the surrogate model has a profound impact on the computational cost and convergence characteristics of surrogate model-based design optimization (Zhou et al. 2016b). The accuracy of surrogate models may vary for different problems. For example, Simpson et al. (2001) found kriging models to be the most accurate for slightly nonlinear responses in high dimensions, whereas Fang et al. (2005) found that RBF models show the best quality for highly nonlinear responses. To avoid a poor-quality surrogate model, the common practice is to construct multiple surrogate models based on input and output data from simulations, evaluate the accuracy of these surrogate models, and then select the single surrogate model with the best quality (Acar and Rais-Rohani 2009). However, this method does not take full advantage of the construction of the different surrogate models, and evaluating the accuracy of multiple surrogate models may consume additional resources. An alternative method to overcome these drawbacks and improve the prediction accuracy of surrogate modelling is to form an ensemble by combining the individual surrogate models. Such an ensemble of surrogate models can take advantage of each individual surrogate model, effectively reducing the error caused by the instability of a single surrogate model. The most common ensemble of surrogate models is a linear combination of individual surrogate models. The key step in constructing an ensemble of surrogate models is the selection of the weight coefficients. Usually, surrogates with higher prediction accuracies have larger weight coefficients, while surrogates with poorer prediction accuracies have smaller weight coefficients. Various methods for selecting the weight coefficients have been studied. For example, Bishop (1995) proposed an optimal weighted ensemble of surrogates in which a matrix method is used to calculate the weight coefficients. Such as weight coefficient selection based on error minimization (Acar and Rais-Rohani 2009). Zerpa (2005) set the predicted variance for local error estimation as the weight coefficient for each surrogate model. Zhou et al. (2011) obtained the optimal average weight by using a recursive algorithm to minimize the predicted root mean square error (RMSE). Details of ensemble surrogate models can be found in Chap. 3.

Simulation models with different fidelities have been widely used in the design of complex systems. Three common ways to obtain a low-fidelity (LF) model are

[43] (a) to simplify the analysis model (e.g. by using a coarse finite element mesh instead of a refined mesh), (b) to simplify the modelling concept or domain (e.g. by using a two-dimensional (2D) model instead of a three-dimensional (3D) model) and (c) to simplify the mathematical or physical description of the system (e.g. by using the Euler non-cohesive equations instead of the Navier–Stokes viscous Newton equations). Generally, sample points from high-fidelity (HF) simulations and models offer more information about the system but at a higher cost (Shan and Wang 2010). It is impractical to directly incorporate HF models into design optimization because the evaluation of a large number of design alternatives would incur a high computational cost. However, optimization based on LF models may lead to a false optimum. A promising way to achieve a trade-off between high accuracy and efficiency is to adopting multi-fidelity (MF) surrogate models (Zhou et al. 2015). The goal of MF surrogate modelling is to combine data obtained from both LF and HF models, with the LF data indicating the trends and a small number of HF simulations for calibrating the LF model. The three main approaches for constructing MF surrogate models are scaling-function-based MF surrogate modelling (Burgee et al. 1996; Sun et al. 2012; Tyan et al. 2015), space mapping (Bandler et al. 2004; Rayas-Sanchez 2016) and co-kriging (Kennedy and O'Hagan 2000; Xiong et al. 2013).

These surrogate modelling techniques play an important role in supporting engineering design and optimization (Shu et al. 2017): (1) engineers can gain insight into a system by employing a cheap-to-run surrogate model, (2) surrogate models offer a better noise filtering capability than gradient-based methods do, (3) building a surrogate model makes it easier to detect simulation errors and identify interesting regions as the entire design space is explored and analysed and (4) building a surrogate model makes parallel computing and optimization simpler because it involves running the same simulation for many design alternatives (Chen et al. 2006). In summary, surrogate modelling is a method that is applied in many disciplines in place of simulation models or physical experiments. It serves as a cheap and powerful tool for computational expensive analysis and design and optimization of complex real-world systems. Much more could be said about surrogate models. There are several other published books on this topic in addition to those mentioned above, e.g. Wan (2009).

1.2 Symbols and Terminology

In this book, several different kinds of variables are used, e.g. scalars, vectors, matrices and random variables. These variables have different representations in different research areas, and there are several custom notations. We will attempt to list some of the notations that are frequently seen so that the reader can understand them more easily.

Usually, non-bold letters are used to represent scalars, and bold letters are used to represent vectors or matrices.

$\widehat{f}(\cdot)$, $\widehat{y}(\cdot)$ and $\widehat{g}(\cdot)$ are often used to denote surrogate models; see Eqs. (3.1) and (5.1).

Greek letters are often used to denote the parameters of surrogate models; e.g. β_i and β_{ij} represent the parameters of a PRS model in Eq. (2.1). Greek letters with hat symbols (e.g. $\hat{\sigma}^2$ and $\hat{\beta}$) denote estimates of these parameters; see Eq. (2.36).

$\boldsymbol{x}$ usually denotes a vector of design variables. y is usually used to denote the response of a design. For multiple vectors of design variables, different subscripts are used to distinguish them.

Certain lowercase letters (i, j, m and n) are often used to denote numbers, e.g. the number of design variables or the number of sample points.

The general rules above cannot fully cover the usage of all symbols. We present a description of the meaning of the symbols in each equation. The reader should infer from the context the type of variable represented by each symbol and its meaning.

References

Acar E, Rais-Rohani M (2009) Ensemble of metamodels with optimized weight factors. Struct Multidiscip Optim 37:279–294

Bandler JW, Cheng QS, Nikolova NK, Ismail MA (2004) Implicit space mapping optimization exploiting preassigned parameters. IEEE Trans Microw Theory Tech 52:378–385

Bishop CM (1995) Neural networks for pattern recognition. Oxford University Press

Burgee S, Giunta AA, Balabanov V, Grossman B, Mason WH, Narducci R, Haftka RT, Watson LT (1996) A coarse-grained parallel variable-complexity multidisciplinary optimization paradigm. Int J Supercomput Appl High Perform Comput 10:269–299

Can B, Heavey C (2012) A comparison of genetic programming and artificial neural networks in metamodeling of discrete-event simulation models. Comput Oper Res 39:424–436

Chen VC, Tsui K-L, Barton RR, Meckesheimer M (2006) A review on design, modeling and applications of computer experiments. IIE Trans 38:273–291

Christelis V, Mantoglou A (2016) Pumping optimization of coastal aquifers assisted by adaptive metamodelling methods and radial basis functions. Water Resour Manage 30:5845–5859

Crombecq K, Laermans E, Dhaene T (2011) Efficient space-filling and non-collapsing sequential design strategies for simulation-based modeling. Eur J Oper Res 214:683–696

Fang H, Horstemeyer MF (2006) Global response approximation with radial basis functions. Eng Optim 38:407–424

Fang H, Rais-Rohani M, Liu Z, Horstemeyer M (2005) A comparative study of metamodeling methods for multiobjective crashworthiness optimization. Comput Struct 83:2121–2136

Kennedy MC, O'Hagan A (2000) Predicting the output from a complex computer code when fast approximations are available. Biometrika 87:1–13

Kleijnen JP (1987) Statistical tools for simulation practitioners. Marcel Dekker

Kleijnen JP (2008) Response surface methodology for constrained simulation optimization: an overview. Simul Model Pract Theory 16:50–64

Kleijnen JP (2009) Kriging metamodeling in simulation: a review. Eur J Oper Res 192:707–716

Rayas-Sanchez JE (2016) Power in simplicity with ASM: tracing the aggressive space mapping algorithm over two decades of development and engineering applications. IEEE Microwave Mag 17:64–76

Razavi S, Tolson BA, Burn DH (2012) Review of surrogate modeling in water resources. Water Resour Res 48

Shan S, Wang GG (2010) Survey of modeling and optimization strategies to solve high-dimensional design problems with computationally-expensive black-box functions. Struct Multidiscip Optim 41:219–241

Shu L, Jiang P, Wan L, Zhou Q, Shao X, Zhang Y (2017) Metamodel-based design optimization employing a novel sequential sampling strategy. Eng Comput 34:2547–2564

Simpson TW, Mauery TM, Korte JJ, Mistree F (2001) Kriging models for global approximation in simulation-based multidisciplinary design optimization. AIAA J 39:2233–2241

Sun G, Li G, Li Q (2012) Variable fidelity design based surrogate and artificial bee colony algorithm for sheet metal forming process. Finite Elem Anal Des 59:76–90

Tyan M, Nguyen NV, Lee J-W (2015) Improving variable-fidelity modelling by exploring global design space and radial basis function networks for aerofoil design. Eng Optim 47:885–908

Van Gelder L, Das P, Janssen H, Roels S (2014) Comparative study of metamodelling techniques in building energy simulation: guidelines for practitioners. Simul Model Pract Theory 49:245–257

Viana FA, Simpson TW, Balabanov V, Toropov V (2014) Special section on multidisciplinary design optimization: metamodeling in multidisciplinary design optimization: how far have we really come? AIAA J 52:670–690

Wan X (2009) Stochastic optimization with simulation based optimization: a surrogate model framework. VDM Publishing

Xiao M, Gao L, Shao X, Qiu H, Jiang P (2012) A generalised collaborative optimisation method and its combination with kriging metamodels for engineering design. J Eng Des 23:379–399

Xiong S, Qian PZ, Wu CJ (2013) Sequential design and analysis of high-accuracy and low-accuracy computer codes. Technometrics 55:37–46

Zerpa LE, Queipo NV, Pintos S, Salager J-L (2005) An optimization methodology of alkaline–surfactant–polymer flooding processes using field scale numerical simulation and multiple surrogates. J Petrol Sci Eng 47:197–208

Zheng J, Shao X, Gao L, Jiang P, Li Z (2013) A hybrid variable-fidelity global approximation modelling method combining tuned radial basis function base and kriging correction. J Eng Des 24:604–622

Zhou XJ, Ma YZ, Li XF (2011) Ensemble of surrogates with recursive arithmetic average. Struct Multidiscip Optim 44:651–671

Zhou Q, Shao X, Jiang P, Cao L, Zhou H, Shu L (2015) Differing mapping using ensemble of metamodels for global variable-fidelity metamodeling. CMES Comput Model Eng Sci 106:323–355

Zhou Q, Shao X, Jiang P, Gao Z, Wang C, Shu L (2016a) An active learning metamodeling approach by sequentially exploiting difference information from variable-fidelity models. Adv Eng Inform 30:283–297

Zhou Q, Shao X, Jiang P, Gao Z, Zhou H, Shu L (2016b) An active learning variable-fidelity metamodelling approach based on ensemble of metamodels and objective-oriented sequential sampling. J Eng Des 27:205–231

Zhou Q, Wang Y, Jiang P, Shao X, Choi S-K, Hu J, Cao L, Meng X (2017) An active learning radial basis function modeling method based on self-organization maps for simulation-based design problems. Knowl Based Syst 131:10–27

Chapter 2
Classic Types of Surrogate Models

2.1 Polynomial Response Surface Models

The polynomial response surface (PRS) methodology is a statistical technique that uses regression analysis and analysis of variance to determine the relationship between design variables and responses. A linear polynomial is used to approximate the implicit limit state equation. The coefficients of the linear polynomial are determined through experimental design. The general form of a PRS model relation is as follows:

$$f(x) = \beta_0 + \sum_{i=1}^{m} \beta_i \boldsymbol{x}_i + \sum_{i=1}^{m} \sum_{j \geq i}^{m} \beta_{ij} \boldsymbol{x}_i \boldsymbol{x}_j + \cdots + \varepsilon \tag{2.1}$$

where ε is the statistical error; $\boldsymbol{x}_i$ is the i-th component of the m-dimensional predictor; and β_0, β_i and β_{ij} are parameters to be estimated and can be arranged in a certain order to form a column vector $\boldsymbol{\beta}$. The key to solving for the fitted model of a polynomial is to solve for the column vector $\boldsymbol{\beta}$. Low-order polynomials are usually selected for response surface models, and second-order PRS models are widely used because of their flexibility and ease of use. The number of parameters to be estimated for such a model is $(m+1)(m+2)/2$, and the number of sample points is usually greater than $(m+1)(m+2)/2$. By substituting the sample point data into Eq. (2.1), the model can be written in matrix product form as

$$f(x^i) = X^i \cdot \boldsymbol{\beta} + \varepsilon, (i = 1, 2, \ldots, n) \tag{2.2}$$

where X^i is the row vector formed by the sample point components $\boldsymbol{x}_i$ in the order of the corresponding components in β, and the following formula needs to be satisfied:

P. Jiang et al., *Surrogate Model-Based Engineering Design and Optimization*,
Springer Tracts in Mechanical Engineering,
https://doi.org/10.1007/978-981-15-0731-1_2

$$f(x^i) = y^i \tag{2.3}$$

where $f(x^i)$ is the predicted value and y^i is the exact value. Thus, we have

$$\mathbf{X} \cdot \boldsymbol{\beta} = \mathbf{Y} \tag{2.4}$$

where $\mathbf{X} = \begin{pmatrix} X^i \\ \vdots \\ X^n \end{pmatrix}$ is a matrix and $\mathbf{Y} = \begin{pmatrix} y^i \\ \vdots \\ y^n \end{pmatrix}$ is a vector. In accordance with the principle of least squares, we can obtain

$$\boldsymbol{\beta} = \left[X^T X\right]^{-1} X^T Y \tag{2.5}$$

By substituting Eq. (2.5) into (2.1), we can obtain the required polynomial fitting model. Its fit function is as follows:

$$\widehat{f}(x_i) = \widehat{\beta_0} + \sum_{i=1}^{m} \widehat{\beta}_i \boldsymbol{x}_i + \sum_{i=1}^{m} \sum_{j \geq i}^{m} \widehat{\beta_{ij}} \boldsymbol{x}_i \boldsymbol{x}_j \tag{2.6}$$

Here, the variables have the same definitions as in Eq. (2.1), and '^' represents an estimated value.

A PRS model has good continuity and conductivity, can effectively suppress the influence of digital noise and is easy to optimize. Moreover, the magnitudes of the coefficients of the components in Eq. (2.1) can be used as a basis to judge the magnitude of the influence of each parameter on the response of the entire system. However, in the case of highly nonlinear high-dimensional problems, the fitting prediction effect of a PRS model is not ideal, and overfitting can occur when the polynomial order is high. These problems are caused by the inability of polynomials to represent highly nonlinear relations. To construct a suitable model using polynomial methods, some suitable function $\psi_i(\boldsymbol{x})$ can be considered in place of $\boldsymbol{x}_i$ in Eq. (2.1). The resulting model can be called a generalized PRS model, and the mathematical form of such a model is

$$f(x) = \sum_{i=0}^{m} \beta_i \psi_i(\boldsymbol{x}) \tag{2.7}$$

where the function $\psi_i(\boldsymbol{x})$ is determined by the model designer in accordance with the physical properties of the problem itself.

After a response surface model is constructed, to ensure the adaptability of the model and determine whether it can truly reflect the statistical relationship between the optimization target and the design variables, its predictive ability must be evaluated. Verification must be performed by means of a deterministic factor test.

The deterministic coefficient is calculated as shown in Eq. (2.8). The value of R^2 is between 0 and 1. The closer the value of R^2 is to 1, the higher the reference value of the related equation; on the contrary, the closer to 0 the deterministic coefficient is, the lower the reference value:

$$R^2 = 1 - \frac{SS_E}{SS_T} \tag{2.8}$$

where the sum of the squared residuals is

$$SS_E = \sum_{i=1}^{n} (y_i - \widehat{y}_i)^2 \tag{2.9}$$

and the total sum of the squares is

$$SS_T = \sum_{i=1}^{n} (y_i - \overline{y}_i)^2 \tag{2.10}$$

2.2 Radial Basis Function Models

A radial function is a function in which the Euclidean distance between the point to be measured and the sample point is used as the independent variable. A model constructed via linear superposition with radial functions as the basis functions is a radial basis function model. Radial basis functions are widely used in many fields, such as discrete data interpolation and image processing.

The basic idea of a radial basis function model is as follows: First, a certain number of sample points $\boldsymbol{x}^i = \{x_1^i, x_2^i, x_3^i, \ldots, x_m^i\}(i = 1, 2, \ldots, n)$ in the design space are selected through experimental design. Then, each of the selected sample points is taken as the centre, and the corresponding radial function is used as a basis function, and these radial functions are linearly fitted to obtain the response value of the point x to be measured. Finally, the Euclidean distance between the sample point and the point to be measured is used as an independent variable to transform a complex multidimensional problem into a simple one-dimensional problem.

When an experimental design yields n sample points, the simple expression for the radial basis function proxy model is

$$f(\boldsymbol{x}) = \sum_{i=1}^{n} \beta_i \cdot \phi(r^i) = \boldsymbol{\beta}^T \boldsymbol{\phi} \tag{2.11}$$

where $\boldsymbol{\beta} = \begin{pmatrix} \beta_1 \\ \vdots \\ \beta_n \end{pmatrix}$, $\phi = \begin{pmatrix} \phi(r^i) \\ \vdots \\ \phi(r^i) \end{pmatrix}$, the β_i are the weight coefficients, $\phi(r)$ is a radial function and $r^i = \|\boldsymbol{x} - \boldsymbol{x}^i\|$ is the Euclidean distance between the point $\boldsymbol{x}$ to be measured and the sample point $\boldsymbol{x}^i$.

When Eq. (2.11) is used as a prediction model, it must satisfy the following interpolation conditions:

$$f(x^j) = y^j, (j = 1, 2, \cdots, n) \tag{2.12}$$

where $f(\boldsymbol{x}^j)$ is the predicted value and y^j is the exact value. Substituting Eq. (2.12) into Eq. (2.11) yields the following formula:

$$\boldsymbol{\Phi} \cdot \boldsymbol{\beta} = \mathbf{Y} \tag{2.13}$$

where $\boldsymbol{\Phi} = \begin{pmatrix} \phi(\|\boldsymbol{x}^1 - \boldsymbol{x}^1\|) & \cdots & \phi(\|\boldsymbol{x}^1 - \boldsymbol{x}^n\| \\ \vdots & \vdots & \vdots \\ \phi(\|\boldsymbol{x}^n - \boldsymbol{x}^1\| & \cdots & \phi(\|\boldsymbol{x}^n - \boldsymbol{x}^n\| \end{pmatrix}$ and $\mathbf{Y} = \begin{pmatrix} y^1 \\ \vdots \\ y^n \end{pmatrix}$.

When the sample points in Eq. (2.8) do not coincide and the matrix $\boldsymbol{\Phi}$ is non-singular, there is a unique solution, and the weight coefficients can be obtained as follows:

$$\boldsymbol{\beta} = \boldsymbol{\Phi}^{-1} \cdot \mathbf{Y} \tag{2.14}$$

The radial functions that are commonly used in radial basis function proxy models are shown in Table 2.1.

In this table, r is the Euclidean distance between the point x to be measured and any sample point, and c is a constant parameter greater than zero, which is also a shape parameter.

The properties of radial basis function models vary with the radial functions employed. When Gaussian functions or inverse multi-quadric functions are used as the basis functions in a radial basis function model, the model will possess characteristics of local estimation due to the influence of the radial function. By contrast,

Table 2.1 Common radial basis functions

Related function name	$\phi(r)$
Gauss function	$\phi(r) = \exp(\frac{-r^2}{c^2}), c > 0$
Multi-quadric (MQ) function	$\phi(r) = \sqrt{r^2 + c^2}, c > 0$
Inverse multi-quadric function	$\phi(r) = \frac{1}{\sqrt{r^2 + c^2}}, c > 0$
Thin-plate splines function	$\phi(r) = (\frac{r}{c^2}) \log(\frac{r}{c}), c > 0$

when a multi-quadric function is used as the kernel function, the model will possess characteristics of global estimation. Radial basis function models are proxy models with good flexibility, simple structures, relatively few calculations and high efficiency. The most commonly used basis function form is a Gaussian kernel function, and the output of a Gaussian radial basis function model can be expressed as follows:

$$f(\boldsymbol{x}) = \sum_{i=1}^{n} \beta_i \cdot \exp\left(\frac{-(r^i)^2}{c^2}\right), \tag{2.15}$$

The variables in this formula have the same definitions as in Eq. (2.11), i.e. $r^i = \|\boldsymbol{x} - \boldsymbol{x}^i\|$ is the Euclidean distance between the point $\boldsymbol{x}$ to be measured and the sample point $\boldsymbol{x}^i$.

2.3 Support Vector Regression Models

A support vector regression (SVR) model is a support vector machine (SVM) model applied to a regression problem. SVR models can be divided into linear SVR models and nonlinear SVR models. Linear SVR models are suitable for simple linearly separable cases, while nonlinear SVR models are suitable for high-dimensional, complex and linearly inseparable cases. The construction process and mechanism of the two types of models are the same; the difference is that in a nonlinear SVR model, the linear segmentation function in a linear SVR model is replaced with a kernel function. Therefore, to clarify the modelling mechanism for SVR models, a linear SVR model will be introduced here to serve as an example.

In the process of constructing a linear SVR model, the goal is to find a linear function $f(\boldsymbol{x})$ to separate the data. $f(\boldsymbol{x})$ can be expressed as follows:

$$f(\boldsymbol{x}) = \langle \omega \cdot \boldsymbol{x} \rangle + b \tag{2.16}$$

Here, $\langle \omega \cdot \boldsymbol{x} \rangle$ denotes the inner product of ω and $\boldsymbol{x}$.

In Fig. 2.1, the red line represents $f(\boldsymbol{x})$, and the points represent all sample points in the training set.

There are two basic requirements for building a linear SVR model: (1) Each input training sample point must deviate from $f(\boldsymbol{x})$ by no more than ε to ensure that $f(\boldsymbol{x})$ is well characterized for the training set. (2) The value of ω in the trained $f(\boldsymbol{x})$ should be as small as possible to ensure the smoothness of the $f(\boldsymbol{x})$ function. The smoother $f(\boldsymbol{x})$ is, the better the prediction performance. In accordance with these two basic requirements, the linear SVR model can be constructed as follows:

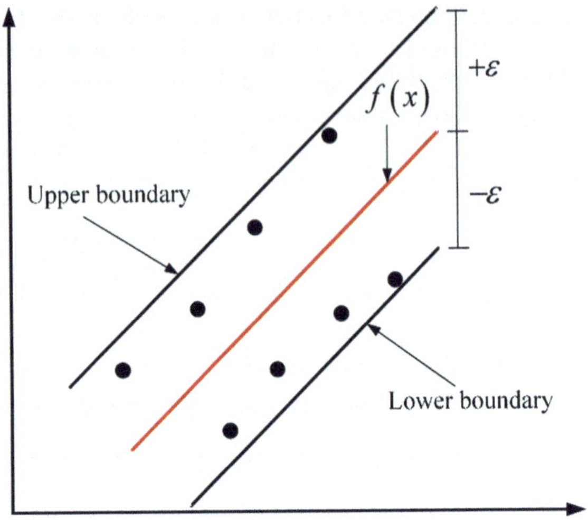

Fig. 2.1 Modelling with a linear SVR model

$$\begin{aligned}&\min \quad \frac{1}{2}|w|^2 \\ &s.t. \quad \begin{cases} y_i - \langle w \cdot \boldsymbol{x} \rangle - b \leq \varepsilon \\ \langle w \cdot \boldsymbol{x} \rangle + b - y_i \leq \varepsilon \end{cases} \quad i = 1, 2, 3, \ldots, n\end{aligned} \tag{2.17}$$

where n denotes the total number of sample points in the training set.

Equation (2.17) is a mathematical model obtained by assuming that the deviation between $f(\boldsymbol{x})$ and each sample point does not exceed ε. However, in a real modelling scenario, there may be intervals in which individual sample points exceed ±ε, and the information contained in these points is valuable for the construction of the model, as shown in Fig. 2.2. To allow some sample points to exceed the interval of ±ε, that is, some out-of-bounds points, it is necessary to modify the process of model construction accordingly.

For such cases, the slack variables ξ_i and ξ_i^* are introduced, and a penalty factor C $(C > 0)$ is defined. The penalty factor is a preselected constant set to balance the smoothness and slack variable size during the construction of the linear SVR model. With these modifications, the expression for the constructed linear SVR model is

$$\begin{aligned}&\min \quad \frac{1}{2}|w|^2 + C\sum_{i=1}^{n}(\xi_i + \xi_i^*) \\ &s.t. \quad \begin{cases} y_i - \langle w \cdot \boldsymbol{x} \rangle - b \leq \varepsilon + \xi_i \\ \langle w \cdot \boldsymbol{x} \rangle + b - y_i \leq \varepsilon + \xi_i^* \quad i = 1, 2, 3, \ldots, n \\ \xi_i, \xi_i^* \geq 0 \end{cases}\end{aligned} \tag{2.18}$$

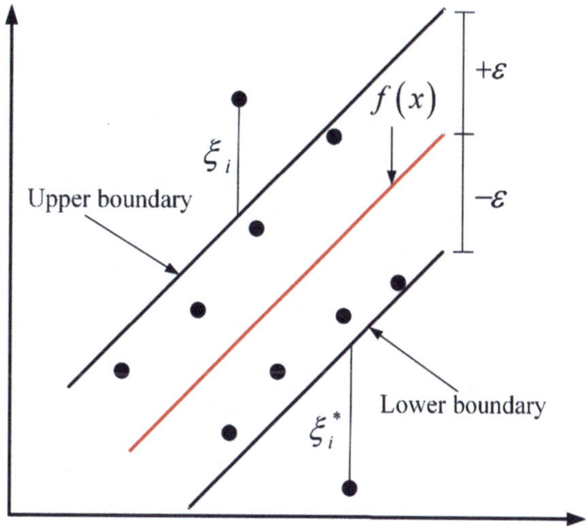

Fig. 2.2 Linear function modelling with a linear SVR model (presenting slack variables)

Lagrangian multipliers are introduced into Eq. (2.18) to solve the optimization problem. The Lagrangian equation is

$$\begin{aligned} L = \frac{1}{2}|w|^2 + C\sum_{i=1}^{n}\left(\xi_i + \xi_i^*\right) - \sum_{i=1}^{n}\alpha_i(\varepsilon + \xi_i - y_i + \langle w \cdot \boldsymbol{x}\rangle + b) \\ - \sum_{i=1}^{n}\alpha_i\left(\varepsilon + \xi_i^* + y_i - \langle w \cdot x\rangle - b\right) - \sum_{i=1}^{n}\left(\beta_i\xi_i + \beta_i^*\xi_i^*\right) \end{aligned} \tag{2.19}$$

In accordance with Lagrangian equation theory, b, w, ξ_i and ξ_i^* in Eq. (2.19) are separately biased to obtain the following equations:

$$\partial_b L = \sum_{i=1}^{n}\left(\alpha_i^* - \alpha_i\right) = 0 \tag{2.20}$$

$$\partial_w L = w - \sum_{i=1}^{n}\left(\alpha_i^* - \alpha_i\right)\boldsymbol{x}_i = 0 \tag{2.21}$$

$$\partial_{\xi_i} L = C - \alpha_i - \eta_i = 0 \tag{2.22}$$

$$\partial_{\xi_i^*} L = C - \alpha_i^* - \eta_i^* = 0 \tag{2.23}$$

By substituting Eqs. (2.20), (2.21), (2.22) and (2.23) into (2.18) along with the slack variables, the original optimization problem can be transformed into the corresponding dual problem as follows:

$$\max\begin{cases} -\frac{1}{2}\sum_{i,j=1}^{n}\left(\alpha_i-\alpha_i^*\right)\left(\alpha_j-\alpha_j^*\right)\left\langle \boldsymbol{x}_i\cdot\boldsymbol{x}_j\right\rangle \\ -\varepsilon\sum_{i=1}^{n}\left(\alpha_i+\alpha_i^*\right)+\sum_{i=1}^{n}y_i\left(\alpha_i-\alpha_i^*\right) \end{cases}$$
$$s.t.\begin{cases} \sum_{i=1}^{n}\left(\alpha_i-\alpha_i^*\right)=0 \\ \alpha_i,\alpha_i^*\in[0,C] \end{cases} \tag{2.24}$$

The final expression for the linear regression function to be solved is

$$f(\boldsymbol{x})=\sum_{i=1}^{n}\left(\alpha_i^*-\alpha_i\right)\left\langle \boldsymbol{x}_i\cdot\boldsymbol{x}\right\rangle+b \tag{2.25}$$

The construction mechanism for a nonlinear SVR model is essentially the same as that for a linear SVR model. During the construction process, the kernel function $k(\boldsymbol{x}_i,\boldsymbol{x}_j)$ is introduced, and the linear indivisibility condition is mapped to a high-dimensional feature space to achieve linear separability in that feature space. The working principle of the kernel function is illustrated in Fig. 2.3.

By introducing the kernel function $k(\boldsymbol{x}_i,\boldsymbol{x}_j)$ into Eq. (2.24), the following mathematical expression is obtained for a nonlinear SVR model:

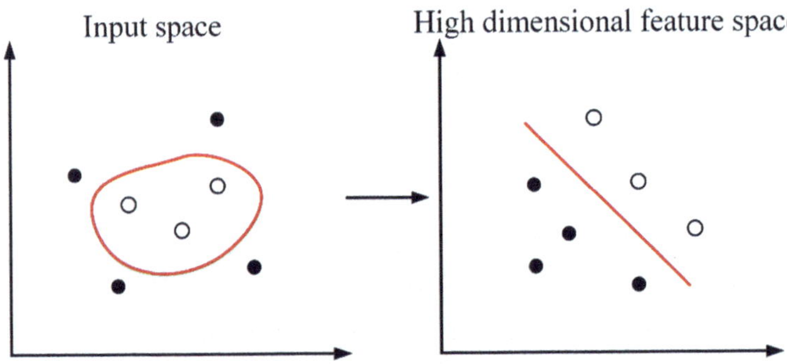

Fig. 2.3 Schematic diagram of the working principle of the kernel function

Table 2.2 Common kernel functions for SVR models

Name	Expression
Linear kernel function	$k(\boldsymbol{x},\boldsymbol{x}') = \boldsymbol{x}^T \cdot \boldsymbol{x}'$
Polynomial kernel function	$k(\boldsymbol{x},\boldsymbol{x}') = \langle \boldsymbol{x} \cdot \boldsymbol{x}' \rangle^d$
Gaussian kernel function	$k(\boldsymbol{x},\boldsymbol{x}') = \exp\left(-\frac{x-x'}{2\sigma^2}\right)$
Sigmoid kernel function	$k(\boldsymbol{x},\boldsymbol{x}') = \tanh(k\langle \boldsymbol{x} \cdot \boldsymbol{x}' \rangle + \theta)$
Non-homogeneous polynomial kernel function	$k(\boldsymbol{x},\boldsymbol{x}') = (\langle \boldsymbol{x} \cdot \boldsymbol{x}' \rangle + c)^d$

$$
\max \begin{cases} -\frac{1}{2}\sum_{i,j=1}^{n}\left(\alpha_i - \alpha_i^*\right)\left(\alpha_j - \alpha_j^*\right)k\langle \boldsymbol{x}_i, \boldsymbol{x}_j\rangle \\ -\varepsilon\sum_{i=1}^{n}\left(\alpha_i + \alpha_i^*\right) + \sum_{i=1}^{n} y_i\left(\alpha_i - \alpha_i^*\right) \end{cases}
$$
$$
s.t. \begin{cases} \sum_{i=1}^{n}\left(\alpha_i - \alpha_i^*\right) = 0 \\ \alpha_i, \alpha_i^* \in [0, C] \end{cases} \tag{2.26}
$$

Therefore, the final nonlinear SVR model is expressed as

$$
f(x) = \sum_{i=1}^{n}\left(\alpha_i^* - \alpha_i\right)k\langle \boldsymbol{x}_i, \boldsymbol{x}\rangle + b \tag{2.27}
$$

Commonly used kernel functions include the linear kernel function, polynomial kernel functions, the Gaussian kernel function, the sigmoid kernel function and non-homogeneous polynomial kernel functions. The specific expressions for these function types are shown in Table 2.2.

2.4 Gaussian Process Models

A Gaussian process (GP) model is an approximate model constructed using a constrained regression method. GP models have been increasingly used in the fields of design optimization and uncertainty quantification for the following reasons (Sasena et al. 2002; Desautels et al. 2014; Zheng et al. 2016): ① GP models belong to the interpolation class of approximation models, that is, a GP model passes through all test sample points, which is necessary for deterministic simulations. ② If there is noise in the output response, a GP model can still be interpolated based on its own modelling mechanism (the 'block gold effect' or error term). ③ Usually, the function to be fitted is a black-box function, that is, the intrinsic properties of this function are unknown; therefore, the fact that GP models offer good adaptability to most function types is advantageous. ④ As a probabilistic model, a GP model is suitable for integrating data from various stages, including a priori information obtained by the designer. ⑤ A GP model can generate prediction errors at non-test-sample points. This feature is especially important for obtaining the prediction accuracy of an approximate model based on a finite number of sample points. In this section, the basic principles of GP models are briefly introduced. For a more detailed theoretical discussion, see the paper by Rasmussen et al. (Audet et al. 2000).

In general, a GP model can be expressed as

$$f(x) \sim GP(m(\cdot), k(\cdot, \cdot)) \tag{2.28}$$

Here, $m(\boldsymbol{x})$ represents the a priori mean function, which can be expressed as

$$m(\boldsymbol{x}) = \boldsymbol{h}(\boldsymbol{x})\boldsymbol{\beta} \tag{2.29}$$

where $\boldsymbol{h}(\boldsymbol{x})$ represents the regression function vector and $\boldsymbol{\beta}$ is the regression coefficient vector.

The expression $k(\boldsymbol{x}, \boldsymbol{x}')$ represents the variance function at design point x and sample point x'; this function can be expressed as

$$k(\boldsymbol{x}, \boldsymbol{x}') = \operatorname{cov}\{f(\boldsymbol{x}), f(\boldsymbol{x}')\} = \sigma^2 \boldsymbol{R}(\boldsymbol{x} - \boldsymbol{x}') \tag{2.30}$$

where $\boldsymbol{X}^s$ is the standard deviation for determining the overall magnitude of the variance, and $\boldsymbol{R}(\boldsymbol{x} - \boldsymbol{x}')$ is the correlation function for the output response at design point $\boldsymbol{x}$ and sample point $\boldsymbol{x}'$. Commonly used forms of correlation functions are shown in Table 2.3.

In Table 2.3, $\theta_1, \ldots, \theta_u$ are roughness parameters indicating the rate at which the correlation between the output responses at design point x and sample point $\boldsymbol{x}'$ decays to zero with the difference between design point x and sample point $\boldsymbol{x}'$.

Here, the Gaussian correlation function is selected to relate the output responses at design point x and sample point $\boldsymbol{x}'$, namely,

$$\boldsymbol{R}(\boldsymbol{x} - \boldsymbol{x}') = \exp\left\{-\sum_{j=1}^{u} \theta_j \left(x_j - x_j'\right)^2\right\} \tag{2.31}$$

Then, the variance function of the output responses at design point $\boldsymbol{x}$ and sample point $\boldsymbol{x}'$ can be expressed as

$$k(\boldsymbol{x}, \boldsymbol{x}') = \sigma^2 \exp\left\{-\sum_{j=1}^{u} \theta_j \left(x_j - x_j'\right)^2\right\} \tag{2.32}$$

Table 2.3 Commonly used correlation function types

Related function name	$\boldsymbol{R}(\boldsymbol{x}-\boldsymbol{x}')$
Exponential function	$\boldsymbol{R}(\boldsymbol{x}-\boldsymbol{x}') = \exp\left\{-\sum_{j=1}^{u}\theta_j/\left\lvert x_j - x_j'\right\rvert\right\}$
Power function	$\boldsymbol{R}(\boldsymbol{x}-\boldsymbol{x}') = \exp\left\{-\sum_{j=1}^{u}\theta_j/\left\lvert x_j - x_j'\right\rvert^2\right\}$
Gaussian function	$\boldsymbol{R}(\boldsymbol{x}-\boldsymbol{x}') = \exp\left\{-\sum_{j=1}^{u}\theta_j\left(x_j - x_j'\right)^2\right\}$
Linear function	$\boldsymbol{R}(\boldsymbol{x}-\boldsymbol{x}') = \max\left\{0, 1-\sum_{j=1}^{u}\theta_j/\left\lvert x_j - x_j'\right\rvert\right\}$
Ball function	$\boldsymbol{R}(\boldsymbol{x}-\boldsymbol{x}') = 1 - 1.5\zeta + 0.5(\zeta)^3$ $\zeta = \min\left\{1, \sum_{j=1}^{u}\theta_j\left\lvert x_j - x_j'\right\rvert\right\}$
Spline function	$\boldsymbol{R}(\boldsymbol{x}-\boldsymbol{x}') = \begin{cases} 1-15(\sum_{j=1}^{u}\theta_j\left\lvert x_j - x_j'\right\rvert)^2 + 30(\sum_{j=1}^{u}\theta_j\left\lvert x_j - x_j'\right\rvert)^3 & 0 \le \sum_{j=1}^{u}\theta_j\left\lvert x_j - x_j'\right\rvert \le 0.2 \\ 1.25(1-(\sum_{j=1}^{u}\theta_j\left\lvert x_j - x_j'\right\rvert)^3) & 0.2 \le \sum_{j=1}^{u}\theta_j\left\lvert x_j - x_j'\right\rvert \le 1 \\ 0 & \sum_{j=1}^{u}\theta_j\left\lvert x_j - x_j'\right\rvert \ge 1 \end{cases}$

where β, σ^2 and θ are hyper-parameters. These hyper-parameters can be obtained through maximum likelihood estimation. The maximum likelihood function of the hyper-parameters can be expressed as follows:

$$p(\mathbf{y}|\boldsymbol{\beta},\sigma^2,\boldsymbol{\theta}) = \frac{1}{(2\pi\sigma^2)^{N/2}|R|^{1/2}}\exp\left\{-\frac{1}{2\sigma^2}(\mathbf{y}-H\boldsymbol{\beta})^T R^{-1}(\mathbf{y}-H\boldsymbol{\beta})^T\right\} \quad (2.33)$$

Taking the logarithmic form of this maximum likelihood function, we obtain

$$\begin{aligned} In(p(\mathbf{y}|\boldsymbol{\beta},\sigma^2,\boldsymbol{\theta})) = & -\frac{N}{2}In(2\pi) - \frac{N}{2}In(\sigma^2) - \frac{1}{2}In(|R|^{1/2}) \\ & -\frac{1}{2\sigma^2}(\mathbf{y}-H\boldsymbol{\beta})^T R^{-1}(\mathbf{y}-H\boldsymbol{\beta})^T \end{aligned} \quad (2.34)$$

To obtain the maximum likelihood estimate of L, the partial derivative of Eq. (2.34) is obtained, and its value is set to 0; thus, the maximum likelihood estimate $\hat{\beta}$ of β is obtained as follows:

$$\hat{\boldsymbol{\beta}} = \left[\mathbf{H}^T\mathbf{R}^{-1}\mathbf{H}\right]^{-1}\mathbf{H}^T\mathbf{R}^{-1}\mathbf{y} \quad (2.35)$$

To obtain the maximum likelihood estimate of σ^2, the maximum likelihood estimate T_u of k is substituted into Eq. (2.34), and the partial derivative of Eq. (2.34) is obtained and set to 0. The maximum likelihood estimate $\hat{\sigma}^2$ of σ^2 is

$$\hat{\sigma}^2 = \frac{1}{N}(\mathbf{y}-H\hat{\boldsymbol{\beta}})^T R^{-1}(\mathbf{y}-H\hat{\boldsymbol{\beta}}) \quad (2.36)$$

Because of the complexity of the mathematical relationship between the maximum likelihood function of the hyper-parameters and $\boldsymbol{\theta}$, it is difficult to obtain the maximum likelihood estimate of $\boldsymbol{\theta}$ by obtaining partial derivatives of the maximum likelihood function as in the cases of $\boldsymbol{x}_i$ and H_l. Therefore, an optimization algorithm is instead used to solve for the maximum likelihood estimate $\hat{\boldsymbol{\theta}}$ of $\boldsymbol{\theta}$. By substituting the maximum likelihood estimates of β and σ^2 into Eq. (2.34), a mathematical model can be established to solve for the maximum likelihood estimate of $\boldsymbol{\theta}$:

$$\max In(p(\mathbf{y}|\boldsymbol{\beta},\sigma^2,\boldsymbol{\theta})) = -\frac{N}{2}In(2\pi) - \frac{N}{2}In(\hat{\sigma}^2) - \frac{1}{2}In(|R|^{1/2}) - \frac{N}{2} \quad (2.37)$$

Here, sequential quadratic programming (SQP) is used to solve the above optimization problem.

Given input sample points $\boldsymbol{x}_o = \left\{x_1^o, x_2^o, \ldots, x_{m_o}^o\right\}$ and the corresponding output responses $\boldsymbol{f}_o = \left\{f_1^o, f_2^o, \ldots, f_{m_o}^o\right\}$, the purpose of establishing a GP model is to estimate the output response $f_p(\boldsymbol{x}_p)$ at design point $\boldsymbol{x}_p = \left\{x_1^p, x_2^p, \ldots, x_{m_p}^p\right\}$. According to the definition of a GP model, the joint Gaussian distribution of the output response $\boldsymbol{f}_o$ at the sample point and the output response $f_p(\boldsymbol{x}_p)$ at the design point can be expressed as

$$\begin{bmatrix} f_o \\ f_p(\boldsymbol{x}_p) \end{bmatrix} \sim N\left(\begin{bmatrix} m(x_o) \\ m(x_p) \end{bmatrix}, \begin{bmatrix} K_o \, K_{op}^T \\ K_{op} \, K_p \end{bmatrix}\right) \tag{2.38}$$

where $m(x_o = H\beta)$ represents the mean vector derived from the sample points. The variance matrix is similar to the variance function $K(x, x')$ and can be decomposed into the training set covariance K_o, the verification set covariance K_p and the training–validation set covariance K_{op}. Since the output response $\boldsymbol{f}_o$ at each sample point is a priori information, the posterior distribution of $f_p(x_p)$ is still a Gaussian distribution, which can be obtained through standard Bayesian analysis:

$$f_p(x_p)|f_o \sim N(E(f_p(x_p)|f_o),\ \mathrm{cov}\left\{f_p(x_p), f_p(x_p')|f_o\right\}) \tag{2.39}$$

In this expression, the mean $E(f_p(x_p)|f_o)$ and the covariance $\mathrm{cov}\{f_p(x_p), f_p(x_p')|f_o\}$ can be calculated as follows:

$$E(f_p(x_p)|f_o) = h(x_p)\hat{\beta} + K_{op}K_o^{-1}(f_o - H\hat{\beta}) \tag{2.40}$$

$$\begin{aligned} \mathrm{cov}\left\{f_p(x_p), f_p(x_p')|f_o\right\} &= K_p - K_{op}K_o^{-1}K_{op}^T \\ &+ (h^T(x) - H^T K_o^{-1}K_{op}^T)^T (H^T K_o^{-1} H)^{-1} (h^T(x) - H^T K_o^{-1} K_{op}^T) \end{aligned} \tag{2.41}$$

Notably, Eq. (2.18) is the posterior distribution of $f_p(x_p)$, and Eq. (2.39) is the prior distribution of $f_p(x_p)$. A numerical example is presented below to illustrate their differences.

$$y = -\sin(x) - e^{x/100} + 10\ ;\ 0 \le x \le 10 \tag{2.42}$$

Figure 2.4 depicts the prior and posterior distributions of $f_p(x_p)$ in the design space. The prior distribution is a GP in which the mean function is a quadratic function. When the information at the five sample points is added to the GP, the posterior distribution of $f_p(x_p)$ in the design space is obtained. The shaded areas in the prior and posterior distributions represent the 95% confidence intervals. Compared with the prior-distribution GP model shown in Fig. 2.4a, the mean and covariance of the posterior-distribution GP model shown in Fig. 2.4b reflect the added sample point information.

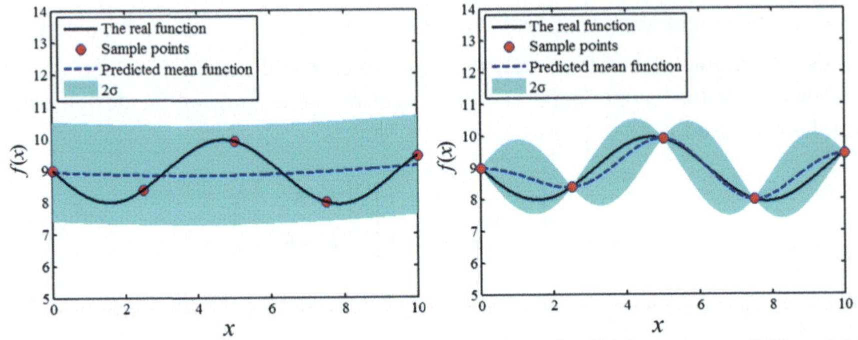

(a) Prior distribution map of GP model (b) Posterior distribution map of GP model

Fig. 2.4 GP model example

2.5 Backpropagation Neural Network Models

A backpropagation (BP) neural network is a multi-layer feedforward network. The gradient descent method is used to train the weight values of such neural networks via backpropagation to solve complex nonlinear problems in engineering. Such models are considered as the starting point of neural network research in the context of modern engineering. BP neural networks are trained through the backward propagation of errors. Error feedback is used to continuously adjust the weight values in the network layer by layer in the reverse direction. Currently, the most widely used neural networks mostly consist of BP networks and their variants. The structure of such a network is shown in Fig. 2.5. When the output error is within the set error range, the number of error steps has exceeded the set value, or a selected maximum number of learning iterations or maximum learning time has been reached, the learning process is terminated and network training ends.

For the three-layer BP neural network shown in Fig. 2.5, the input vector is denoted by $\mathbf{X} = (x_1, x_2, \ldots, x_m)$. The hidden-layer input vector is

Fig. 2.5 Three-layer BP neural network topology

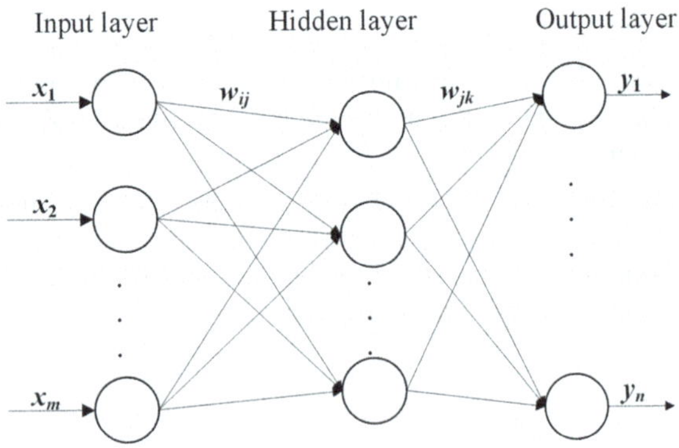

$\mathbf{B} = (b_1, b_2, \ldots b_p)$, the hidden-layer output vector is $\mathbf{C} = (c_1, c_2, \ldots c_p)$, the output-layer input vector is $\mathbf{L} = (l_1, l_2, \ldots l_n)$, the output-layer output vector is $\mathbf{O} = (o_1, o_2, \ldots o_n)$ and the expected final output vector is $\mathbf{Y} = (y_1, y_2, \ldots, y_n)$.

In a BP neural network, the neurons in each layer are connected to the neurons in the next layer in a fully connected manner, that is, each neuron of the current layer is connected to each neuron of the next layer. However, there is no connection relationship between neurons at the same level. The connections between neurons are modified by weights. Therefore, the training process for a BP neural network is actually a process of iteratively updating the connection parameters for the entire network.

Each neuron in a BP neural network is divided into an input and an output, and each neuron has an activation function. The input to the neuron needs to be processed by the activation function to produce an output. The activation functions of all neurons in the same layer are generally the same, whereas different activation functions may be chosen for neurons in different layers. The input layer is generally used only for signal transfer; thus, a linear function can generally be directly used in this layer. In a single-layer perceptual network (M-P model), an early type of neural network, a hidden layer is added to handle nonlinear problems. For the neurons in this layer, the sigmoid function is generally adopted, as shown in Eq. (2.43):

$$S(x) = \frac{1}{1 + e^{-x}} \tag{2.43}$$

The sigmoid function provides the neural network with the ability to solve nonlinear problems. It is characterized by a limited output range of (0, 1). It can be used to predict the probability of a certain sample belonging to a certain category when used to solve a classification problem. In addition, this function is differentiable everywhere in the domain on which it is defined, and its derivative is

$$S'(x) = \frac{e^{-x}}{(1 + e^{-x})^2} = \frac{1}{1 + e^{-x}} - \frac{1}{(1 + e^{-x})^2} = S(x)[1 - S(x)] \tag{2.44}$$

In the case of regression prediction problems, a linear function is generally used as the activation function for the neurons in the output layer. In the case of classification problems, the sigmoid function is generally used in the output layer, in the same way as for the neurons in the hidden layer, and the output value of each neuron is taken as the probability of the corresponding category to serve as a reference for the probability decision. Here, the sigmoid function is used as the excitation function $f(x)$ for both the hidden layer and the output layer.

The steps of training a BP neural network are as follows:

Step 1: Neural network initialization. The connection weights and thresholds of the input layer, the hidden layer and the output layer are randomly assigned values in the range of (−1, 1).

Step 2: The t-th input sample data $X^t = (x_1^t, x_2^t, \ldots, x_m^t)$ and the corresponding expected output sample data $Y^t = (y_1^t, y_2^t, \ldots, y_n^t)$ are extracted using a random method and fed to the neural network.

Step 3: The input b_j and output c_k of each neuron in the hidden layer and the input l_k and output o_k of the output layer are calculated using the connection weights and thresholds of each layer:

$$b_j = \sum_{i=1}^{m} w_{ji} x_i - \theta_j, \qquad j = 1, 2, \ldots, p \tag{2.45}$$

$$c_j = f(b_j), \qquad j = 1, 2, \ldots, p \tag{2.46}$$

$$l_k = \sum_{j=1}^{p} w_{kj} c_j - \theta_k, \qquad k = 1, 2, \ldots, n \tag{2.47}$$

$$o_k = f(l_k), \qquad k = 1, 2, \ldots, n \tag{2.48}$$

Here, w_{ji} is the weight of a connection from the input layer to the hidden layer, that is, the weight of the connection from the i-th neuron of the first layer to the j-th neuron of the second layer. w_{kj} is the weight of a connection from the hidden layer to the output layer, that is, the weight of the connection from the j-th neuron of the second layer to the k-th neuron of the third layer. x_i represents the input to the i-th node of the input layer, o_k represents the input to the k-th node of the input layer, θ_j represents the threshold for the j-th node of the hidden layer, and θ_k represents the threshold for the k-th node of the hidden layer.

Step 4: Using the expected and actual outputs of the network, the training error E^t for the t-th sample is calculated using the established empirical formula:

$$E^t = \frac{1}{2} \sum_{k=1}^{n} (y_k^t - o_k^t)^2 \tag{2.49}$$

Step 5: The error E^t of each neuron obtained from the output layer and the output of each neuron in the hidden layer are used to correct the connection weights and thresholds between the hidden and output layers.

The following formula is used to continuously improve the weights of the output layer:

$$\Delta w_{kj} = -\eta \frac{\partial E^t}{\partial w_{kj}} \tag{2.50}$$

where η is called the learning efficiency of the neural network.

$$\frac{\partial E^t}{\partial w_{kj}} = \frac{\frac{1}{2}\sum_{k=1}^{n}(y_k^t - o_k^t)^2}{\partial w_{kj}} = \frac{\frac{1}{2}\sum_{k=1}^{n}(y_k^t - o_k^t)^2}{\partial o_k}\frac{\partial o_k}{\partial l_k}\frac{\partial l_k}{\partial w_{kj}} \\ = -(y_k^t - o_k^t)f'(l_k)c_j \tag{2.51}$$

where $f\prime(l_k)$ is the partial derivative of the excitation function with respect to the input l_k to the output layer.

Thus, the iterative formula for adjusting the weights of the connections from the hidden layer to the output layer is

$$w_{kj}(t+1) = w_{kj}(t) + \Delta w_{kj} = w_{kj}(t) + \eta(y_k^t - o_k^t)f\prime(l_k)c_j \tag{2.52}$$

The following expression is used to continuously improve the output-layer thresholds:

$$\Delta\theta_k = -\eta\frac{\partial E^t}{\partial\theta_k} \tag{2.53}$$

$$\frac{\partial E^t}{\partial\theta_k} = \frac{\frac{1}{2}\sum\limits_{k=1}^{n}(y_k^t - o_k^t)^2}{\partial\theta_k} = \frac{\frac{1}{2}\sum\limits_{k=1}^{n}(y_k^t - o_k^t)^2}{\partial o_k}\frac{\partial o_k}{\partial l_k}\frac{\partial l_k}{\partial\theta_k} \\ = -(y_k^t - o_k^t)f\prime(l_k)\cdot(-1) = (y_k^t - o_k^t)f\prime(l_k) \tag{2.54}$$

Thus, the iterative formula for adjusting the thresholds between the hidden layer and the output layer is

$$\theta_k(t+1) = \theta_k(t) + \Delta\theta_k = \theta_k(t) - \eta(y_k^t - o_k^t)f\prime(l_k) \tag{2.55}$$

Step 6: The connection weights and thresholds between the input and hidden layers are corrected based on the error of each node of the hidden layer and the input to each neuron of the input layer.

Similarly, the iterative formula for adjusting the weights of the connections from the input layer to the hidden layer is

$$\frac{\partial E^t}{\partial w_{ji}} = \frac{\frac{1}{2}\sum_{k=1}^{n}(y_k^t - o_k^t)^2}{\partial w_{ji}} = \frac{\frac{1}{2}\sum_{k=1}^{n}(y_k^t - o_k^t)^2}{\partial c_j}\frac{\partial c_j}{\partial b_j}\frac{\partial b_j}{\partial w_{ji}} \tag{2.56}$$

where $\frac{\partial E^t}{\partial c_j} = \frac{\sum\limits_{k=1}^{n}(y_k^t - o_k^t)^2}{\partial o_k}\frac{\partial o_k}{\partial l_k}\frac{\partial l_k}{\partial c_j} = -(y_k^t - o_k^t)\cdot f\prime(l_k)\cdot w_{kj}$, $\frac{\partial c_j}{\partial b_j} = f\prime(b_j)$, and $\frac{\partial b_j}{\partial w_{ji}} = x_i$.

Then,

$$\frac{\partial E^t}{\partial w_{ji}} = \frac{\frac{1}{2}\sum_{k=1}^{n}(y_k^t - o_k^t)^2}{\partial w_{ji}} = -(y_k^t - o_k^t)\cdot f\prime(l_k)\cdot w_{kj}\cdot f\prime(b_j)\cdot x_i \tag{2.57}$$

Therefore, the iterative formula for adjusting the weights between the input and hidden layers is

$$w_{ji}(t+1) = w_{ji}(t) + \Delta w_{ji} = w_{ji}(t) + \eta(y_k^t - o_k^t)f'(l_k)w_{kj}f\prime(b_j)x_i \tag{2.58}$$

The expression for the constant improvement of the hidden-layer thresholds is

$$\frac{\partial E^t}{\partial \theta_j} = \frac{\frac{1}{2}\sum_{k=1}^{n}(y_k^t - o_k^t)^2}{\partial \theta_j} = \frac{\frac{1}{2}\sum_{k=1}^{n}(y_k^t - o_k^t)^2}{\partial c_j}\frac{\partial c_j}{\partial b_j}\frac{\partial b_j}{\partial \theta_j} \tag{2.59}$$

Thus,

$$\frac{\partial E^t}{\partial \theta_j} = \frac{\frac{1}{2}\sum_{k=1}^{n}(y_k^t - o_k^t)^2}{\partial \theta_j} = -(y_k^t - o_k^t) \cdot f'(l_k) \cdot w_{kj} \cdot f'(b_j) \cdot (-1) \tag{2.60}$$

Therefore, the iterative formula for adjusting the thresholds between the input layer and the hidden layer is

$$\theta_j(t+1) = \theta_j(t) + \Delta\theta_j = \theta_j(t) - \eta(y_k^t - o_k^t)f'(l_k)w_{kj}f'(b_j) \tag{2.61}$$

Step 7: Another sample is randomly selected from among the remaining training samples to further train the network, and the training process returns to Step 3 until training with N training samples has been completed. The global error E is calculated as follows:

$$E = \frac{1}{N}\sum_{t=1}^{N} E^t \tag{2.62}$$

For the training of the neural network, a maximum number of iterations, a maximum error threshold and a maximum training time are set. When any of these conditions is met, the training ends, and by comparing the training time, the number of training iterations, and the training error, it is judged whether the trained neural network meets the established requirements.

The following figure shows the specific calculation flowchart for a BP neural network.

2.6 Performance Comparison of Different Surrogate Models

In this section, the performance of PRS model, RBF model, SVR model, GP model and BP neural network model is illustrated through eight numerical examples. Some factors that may influence the evaluation results are taken into consideration, such as the dimensions of the problem, number of samples and noise level. Some general conclusions can be obtained from the comparison results, which can be instructions of the selection of surrogate models for problems with different characteristics.

Eight numerical examples, with dimensions from 2 to 16, are used to test these surrogate models. The formulas of these eight problems can be depicted as follows:

Function 1 (F1): SC function

$$\begin{aligned} f(\boldsymbol{x}) =& 4x_1^2 - 2.1x_1^4 + x_1^6/3 + x_1x_2 - 4x_2^2 + 4x_2^4 \\ & \boldsymbol{x} \in [-2, 2]^2 \end{aligned} \tag{2.63}$$

Function 2 (F2): RB function

$$\begin{aligned} f(\boldsymbol{x}) =& 100(x_2 - x_1^2)^2 + (x_1 - 1)^2 \\ & \boldsymbol{x} \in [-2, 2]^2 \end{aligned} \tag{2.64}$$

Function 3 (F3): LF3 function

$$\begin{aligned} f(\boldsymbol{x}) =& \sin^2(\pi\omega_1) + \sum_{i=1}^{2}(\omega_i - 1)^2[1 + 10\sin^2(\pi\omega_i + 1)] \\ & + (\omega_3 - 1)^2[1 + \sin^2(2\pi\omega_3)] \\ \omega_i =& 1 + \frac{x_i - 1}{4}, i = 1, 2, 3, x_i \in [-10, 10] \end{aligned} \tag{2.65}$$

Function 4 (F4): AF4 function

$$\begin{aligned} f(\boldsymbol{x}) =& -20\exp\left(-0.2\sqrt{\frac{1}{n}\sum_{i=1}^{n}x_i^2}\right) - \exp\left(\frac{1}{n}\sum_{i=1}^{n}\cos(2\pi x_i)\right) + 20 + \exp(1) \\ & \boldsymbol{x} \in [-2, 2]^4 \end{aligned} \tag{2.66}$$

Function 5 (F5): HN6 function

$$f(\boldsymbol{x}) = -\frac{1}{1.94}\left[2.58 + \sum_{i=1}^{4} \alpha_i \exp\left(-\sum_{j=1}^{6} A_{ij}(x_j - P_{ij})^2\right)\right], x_{1,\ldots,6} \in (0,1)$$

$$\alpha = (1.0, 1.2, 3.0, 3.2)^T$$

$$\mathbf{A} = \begin{pmatrix} 10 & 3 & 17 & 3.50 & 1.7 & 8 \\ 0.05 & 10 & 17 & 0.1 & 8 & 14 \\ 3 & 3.5 & 1.7 & 10 & 17 & 8 \\ 17 & 8 & 0.05 & 10 & 0.1 & 14 \end{pmatrix} \tag{2.67}$$

$$\mathbf{P} = 10^{-4}\begin{pmatrix} 1312 & 1696 & 5569 & 124 & 8283 & 5886 \\ 2329 & 4135 & 8307 & 3736 & 1004 & 9991 \\ 2348 & 1451 & 3522 & 2883 & 3047 & 6650 \\ 4047 & 8828 & 8732 & 5743 & 1091 & 381 \end{pmatrix}$$

$$\boldsymbol{x} \in [0,1]^6$$

Function 6 (F6): GN8 function

$$f(\boldsymbol{x}) = \frac{1}{4000}\sum_{i=1}^{n} x_i^2 - \prod_{i=1}^{n} \cos\left(\frac{x_i}{\sqrt{i}}\right) + 1 \tag{2.68}$$

$$\boldsymbol{x} \in [-600, 600]^8$$

Function 7 (F7): TR10 function

$$f(\boldsymbol{x}) = \sum_{i=1}^{n} i x_i^2 \tag{2.69}$$

$$\boldsymbol{x} \in [-100, 100]^{10}$$

Function 8 (F8): F16 function

$$f(\boldsymbol{x}) = \sum_{i=1}^{n}\sum_{j=1}^{n} \alpha_{ij}(x_i^2 + x_i + 1)(x_j^2 + x_j + 1), \quad i,j = 1,2,\ldots,n$$

$$\alpha = \begin{bmatrix} 1 & 0 & 0 & 1 & 0 & 0 & 1 & 1 & 0 & 0 & 0 & 0 & 0 & 0 & 0 & 1 \\ 0 & 1 & 1 & 0 & 0 & 0 & 1 & 0 & 0 & 1 & 0 & 0 & 0 & 0 & 0 & 0 \\ 0 & 0 & 1 & 0 & 0 & 0 & 1 & 0 & 1 & 1 & 0 & 0 & 0 & 1 & 0 & 0 \\ 0 & 0 & 0 & 1 & 0 & 0 & 1 & 0 & 0 & 0 & 1 & 0 & 0 & 0 & 1 & 0 \\ 0 & 0 & 0 & 0 & 1 & 1 & 0 & 0 & 0 & 1 & 0 & 1 & 0 & 0 & 0 & 1 \\ 0 & 0 & 0 & 0 & 0 & 1 & 0 & 1 & 0 & 0 & 0 & 0 & 0 & 0 & 1 & 0 \\ 0 & 0 & 0 & 0 & 0 & 0 & 1 & 0 & 0 & 0 & 1 & 0 & 1 & 0 & 0 & 0 \\ 0 & 0 & 0 & 0 & 0 & 0 & 0 & 1 & 0 & 1 & 0 & 0 & 0 & 0 & 1 & 0 \\ 0 & 0 & 0 & 0 & 0 & 0 & 0 & 0 & 1 & 0 & 0 & 1 & 0 & 0 & 0 & 1 \\ 0 & 0 & 0 & 0 & 0 & 0 & 0 & 0 & 0 & 1 & 0 & 0 & 0 & 1 & 0 & 0 \\ 0 & 0 & 0 & 0 & 0 & 0 & 0 & 0 & 0 & 0 & 1 & 0 & 1 & 0 & 0 & 0 \\ 0 & 0 & 0 & 0 & 0 & 0 & 0 & 0 & 0 & 0 & 0 & 1 & 0 & 1 & 0 & 0 \\ 0 & 0 & 0 & 0 & 0 & 0 & 0 & 0 & 0 & 0 & 0 & 0 & 1 & 1 & 0 & 0 \\ 0 & 0 & 0 & 0 & 0 & 0 & 0 & 0 & 0 & 0 & 0 & 0 & 0 & 1 & 0 & 0 \\ 0 & 0 & 0 & 0 & 0 & 0 & 0 & 0 & 0 & 0 & 0 & 0 & 0 & 0 & 1 & 0 \\ 0 & 0 & 0 & 0 & 0 & 0 & 0 & 0 & 0 & 0 & 0 & 0 & 0 & 0 & 0 & 1 \end{bmatrix} \tag{2.70}$$

$$\boldsymbol{x} \in [-1, 1]^{16}$$

Among them, Functions 1–4 are regarded as low-dimensional problems, while Functions 5–8 are regarded as high-dimensional problems (Fig. 2.6).

2.6.1 Influence of Sample Size

The influence of sample size on the performance of different surrogate models is investigated first. Two commonly used metrics, i.e. max absolute error (MAE) and root mean square error (RMSE), are calculated to measure the accuracy of surrogate models (Shu et al. 2017). The MAE evaluates the local accuracy of the surrogate models while the RMSE evaluates the global accuracy. The lower the value of RMSE and MAE, the more accurate the surrogate model. The two error metrics are calculated as

$$RMSE = \sqrt{\frac{1}{m}\sum_{1}^{m} [\hat{y}(x_i) - y(x_i)]^2} \tag{2.71}$$

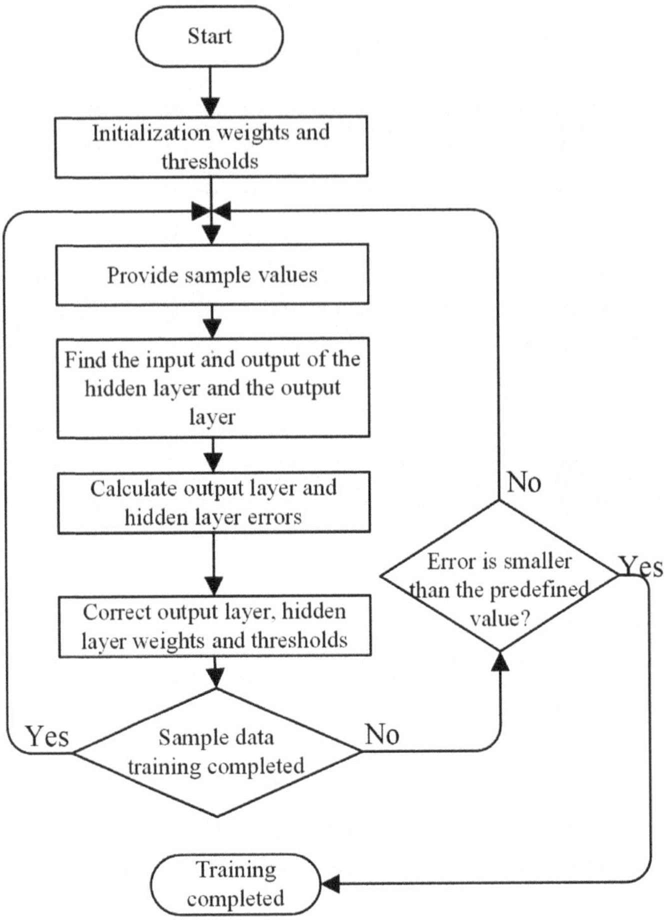

Fig. 2.6 Construction process of the BP neural network

$$MAE = \max|\hat{y}(x_i) - y(x_i)| \tag{2.72}$$

The error metrics of low-dimensional problems obtained by different surrogate models are plotted in Fig. 2.7a–h. The number of sample points is varied from 3 to 11 times of the number of design variables. In addition to the PRS model, it is obvious that with the increasing number of sample points, the prediction error of surrogate models decreases. After the surrogate models achieve a high accuracy, continue adding more sample point will contribute little to the prediction accuracy. The performance of the PRS model is the most unstable. For Function 1 and Function 2, it gives the best accuracy. However, for Function 3 and Function 4, it has the worst prediction performance. The performance of kriging model and RBF model is very stable and better than that of BPNN model and SVR model. In summary, the kriging model and the RBF model are more suitable than other three surrogate models for low-dimensional problems.

The error metrics of high-dimensional problems obtained by different surrogate models are plotted in Fig. 2.8a–h. It can be seen in Fig. 2.8 that the prediction performance of PRS model is worse than the other surrogate models. In addition to the last function, kriging models are the most accurate. For Function 8, the prediction performance of the kriging model is slightly worse than the RBF model but better than other surrogate models. For Function 5–7, the prediction error of the RBF model is not satisfactory. In summary, kriging model is the best choice for high-dimensional problems.

2.6.2 *Influence of Noise Level*

The sampling noise may be included in the responses in engineering applications, which may affect the accuracy of the surrogate model. To test the predictive performance of different surrogate modelling approaches in the presence of noise, artificial noises are added to the response values of the output parameter according to the following formula (Zhao and Xue 2010; Zhou et al. 2016):

$$Y = f(x) + l'\delta \tag{2.73}$$

where $l' = 0$–20% is a scaling parameter and δ is a random number sampled from the standard Gaussian distribution $N \sim (0, 1)$. Five levels of artificial noises, 0%, 5%, 10%, 15% and 20%, are added to the response values of the corresponding sample points using Eq. (2.73). The number of sample points is 10 times of the number of design variables. The accuracy results under different levels of noises are plotted in Figs. 2.9 and 2.10.

Overall, as the noise level increases, the prediction error will increase. In terms of the prediction performance of different surrogate models, PRS model can obtain accurate surrogate model for Function 1 and Function 2, while it shows the worst performance for other functions. For most cases, the prediction error of BPNN model is larger than those of kriging model, RBF model and SVR model. The SVR model does not show particularly good performance for all functions, and it is the most inaccurate for Function 1 and Function 2. The RBF model performs very well only for Function 8. The kriging model is the most accurate surrogate model for almost all test functions; it is only second to the RBF model for Function 8.

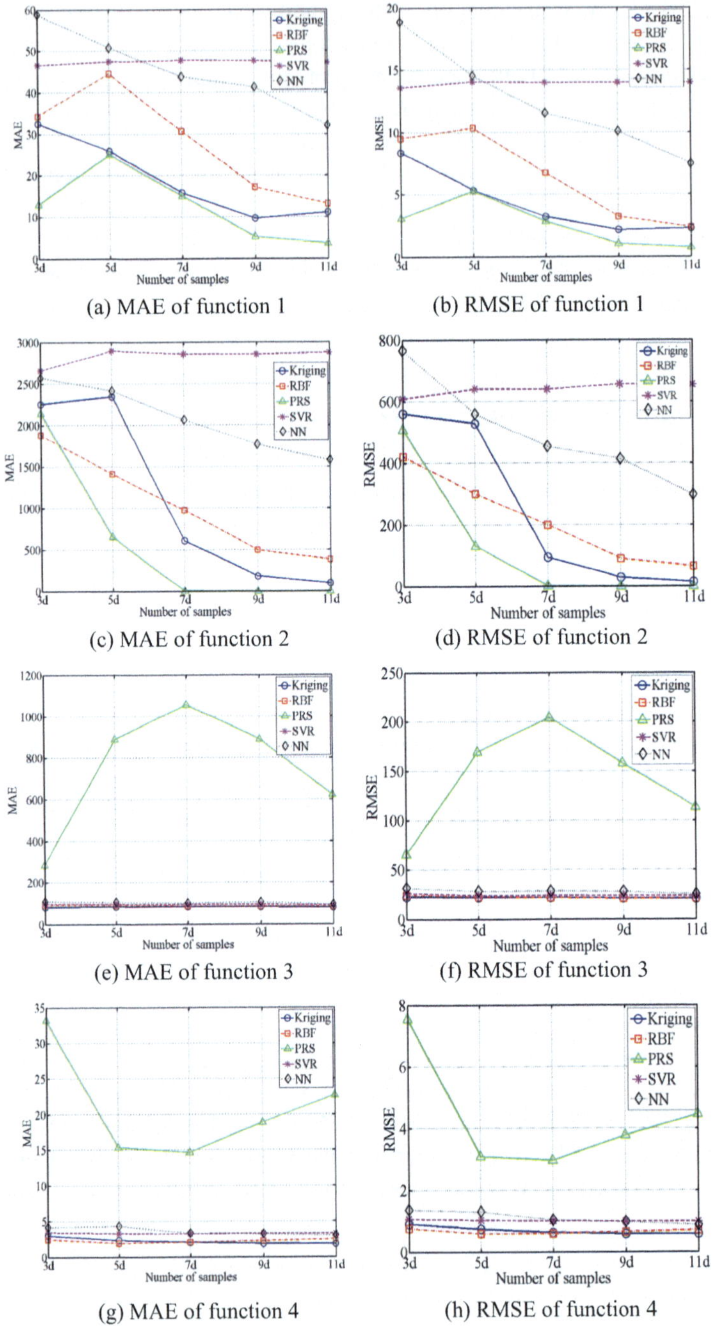

(a) MAE of function 1 (b) RMSE of function 1

(c) MAE of function 2 (d) RMSE of function 2

(e) MAE of function 3 (f) RMSE of function 3

(g) MAE of function 4 (h) RMSE of function 4

Fig. 2.7 Test results of low-dimensional functions under different sample sizes

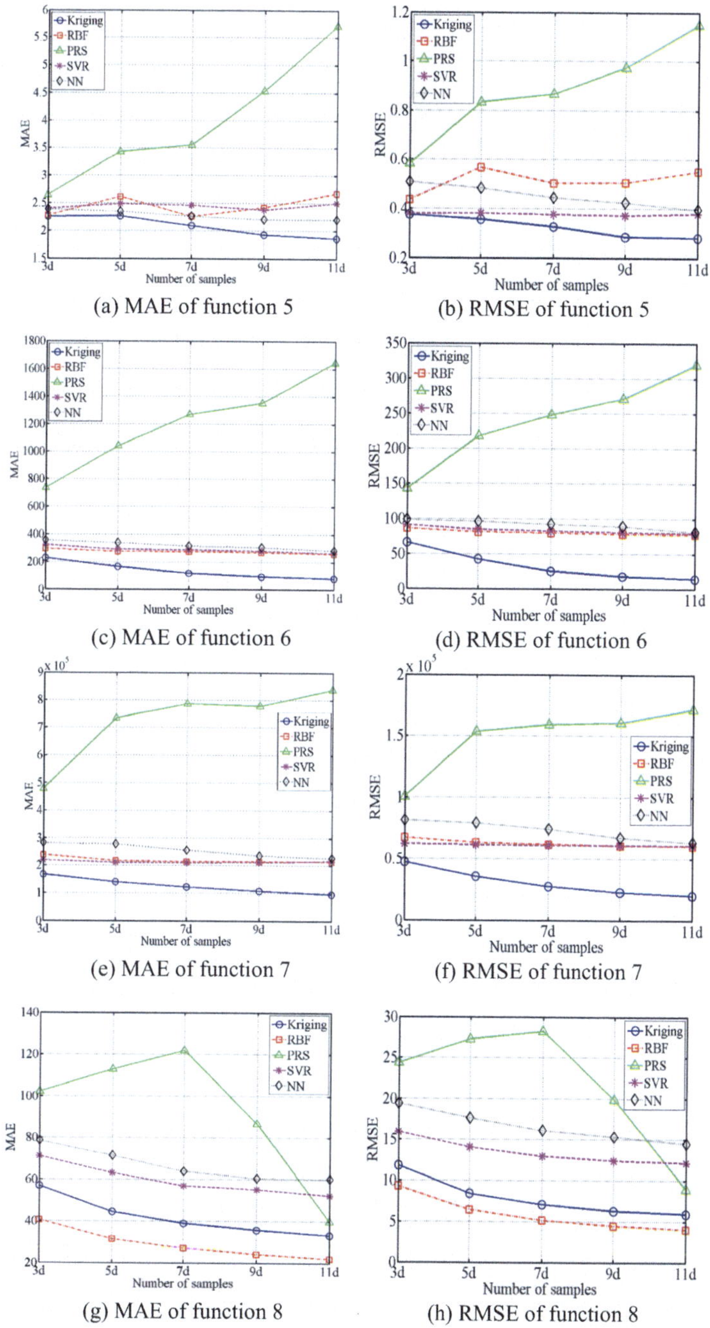

(a) MAE of function 5
(b) RMSE of function 5
(c) MAE of function 6
(d) RMSE of function 6
(e) MAE of function 7
(f) RMSE of function 7
(g) MAE of function 8
(h) RMSE of function 8

Fig. 2.8 Test results of high-dimensional functions under different sample sizes

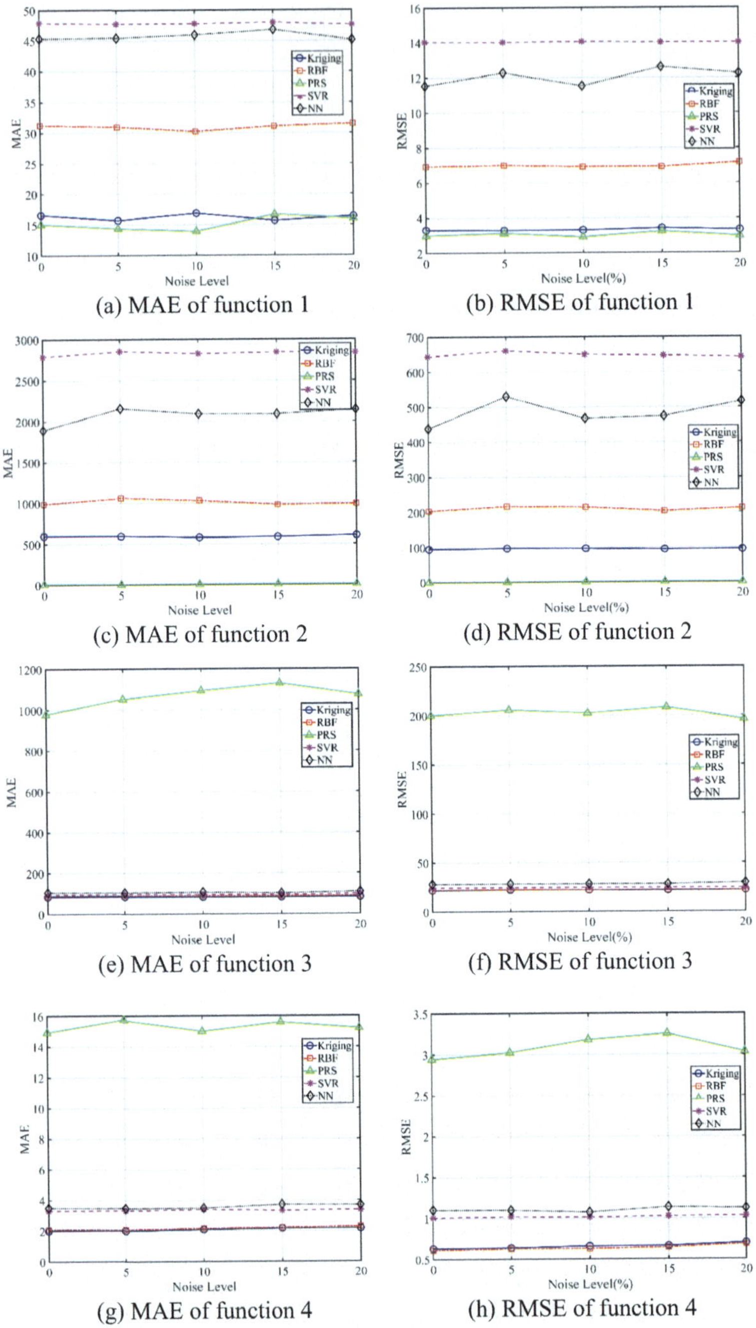

(a) MAE of function 1 (b) RMSE of function 1
(c) MAE of function 2 (d) RMSE of function 2
(e) MAE of function 3 (f) RMSE of function 3
(g) MAE of function 4 (h) RMSE of function 4

Fig. 2.9 Test results of low-dimensional functions under different noise levels

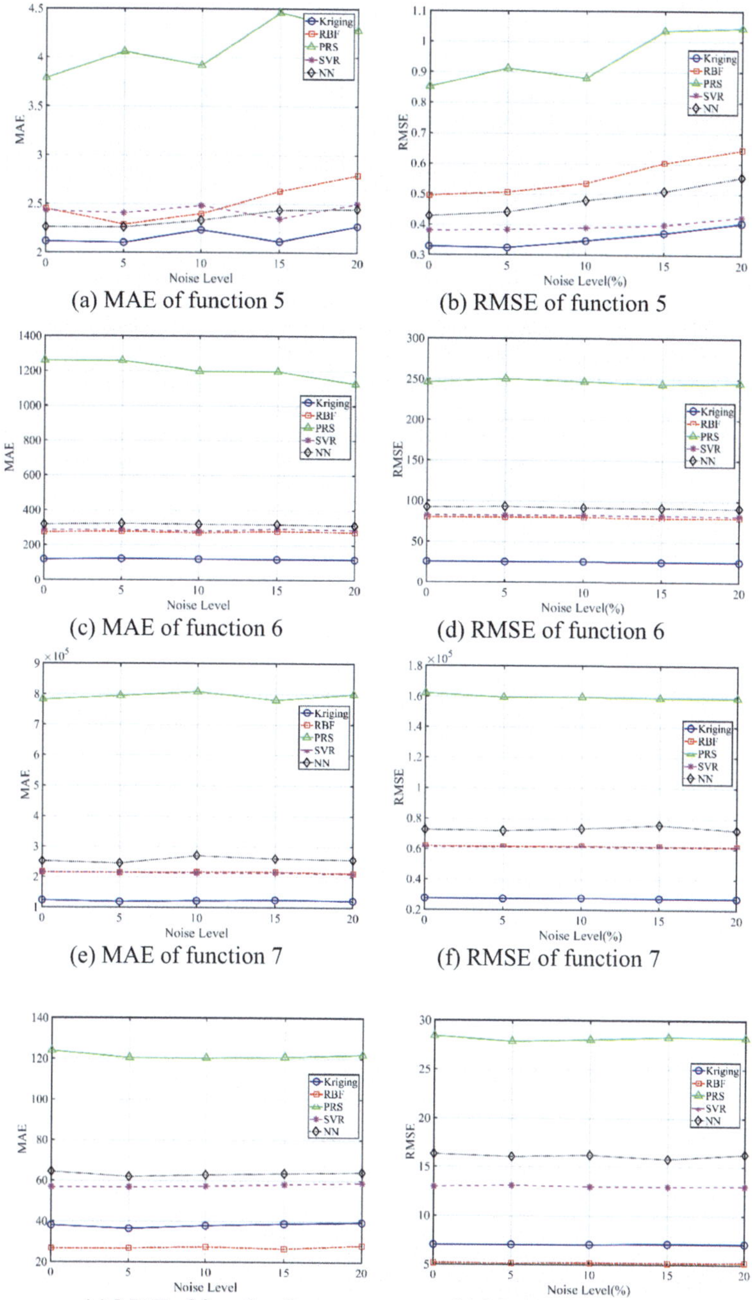

(a) MAE of function 5
(b) RMSE of function 5
(c) MAE of function 6
(d) RMSE of function 6
(e) MAE of function 7
(f) RMSE of function 7
(g) MAE of function 8
(h) RMSE of function 8

Fig. 2.10 Test results of high-dimensional functions under different noise levels

References

Audet C, Denni J, Moore D, Booker A, Frank P (2000) A surrogate-model-based method for constrained optimization. In: 8th symposium on multidisciplinary analysis and optimization, p 4891

Desautels T, Krause A, Burdick J (2014) Parallelizing exploration-exploitation tradeoffs in Gaussian process bandit optimization. 15:3873–3923

Sasena MJ, Papalambros P, Goovaerts P (2002) Exploration of metamodeling sampling criteria for constrained global optimization. 34:263–278

Shu L, Jiang P, Wan L, Zhou Q, Shao X, Zhang Y (2017) Metamodel-based design optimization employing a novel sequential sampling strategy. Eng Comput 34:2547–2564

Zhao D, Xue D (2010) A comparative study of metamodeling methods considering sample quality merits. Struct Multidiscip Optim 42:923–938

Zheng J, Li Z, Gao L, Jiang G (2016) A parameterized lower confidence bounding scheme for adaptive metamodel-based design optimization. 33:2165–2184

Zhou Q, Shao X, Jiang P, Gao Z, Zhou H, Shu L (2016) An active learning variable-fidelity metamodelling approach based on ensemble of metamodels and objective-oriented sequential sampling. J Eng Des 27:205–231

Chapter 3
Ensembles of Surrogate Models

An ensemble of surrogate models (EM) is a surrogate model composed of a series of surrogate models combined through a weighted sum. An EM can take advantage of each individual surrogate model to effectively increase the robustness of the prediction. The mathematical expression for an EM can be given as follows:

$$\hat{y}_e(\boldsymbol{x}) = \sum_{i=1}^{N} w_i(\boldsymbol{x})\hat{y}_i(\boldsymbol{x}) \tag{3.1}$$

where $\boldsymbol{x}$ is the vector of design variables, $\hat{y}_e(\cdot)$ is the response predicted by the EM, N is the number of individual surrogate models, $w_i(\boldsymbol{x})$ is the weight for the i-th surrogate model at point $\boldsymbol{x}$ and $\hat{y}_i(\cdot)$ is the response predicted by the i-th surrogate. The sum of the weights in the EM must be equal to one, that is,

$$\sum_{i=1}^{N} w_i(\boldsymbol{x}) = 1 \tag{3.2}$$

The key step in constructing an EM is to calculate the weight for each surrogate model. Generally, surrogate models with higher prediction accuracies tend to have larger weight coefficients.

The existing EMs can be divided into two categories according to the method used to calculate the weights: weighted average surrogate models and pointwise weighted EMs. The main principle used to calculate the weight coefficients in a weighted average surrogate model is to assign a weight to each component surrogate model in accordance with its global performance. Since global error metrics estimate the errors of each surrogate model over the whole design space, the weight factor for a particular model will be the same at all points. Instead of utilizing global error metrics, the weights for the individual surrogate models in a pointwise weighted EM are calculated using local error metrics, which means that the weight for each individual surrogate model will vary throughout the design space.

P. Jiang et al., *Surrogate Model-Based Engineering Design and Optimization*, Springer Tracts in Mechanical Engineering,
https://doi.org/10.1007/978-981-15-0731-1_3

Compared with the average weighting method, the pointwise weighting strategy can better capture the local features of the objective function by allowing flexible adjustments of the local weight coefficients in the design space.

Some typical methods of creating an EM are introduced in this chapter.

3.1 Weighted Average Surrogate Models

3.1.1 Weight Factor Selection Based on the Generalized Mean Square Cross-Validation Error (GMSE)

Cross-validation errors are widely used in weight selection for EMs. The sample points are divided into a training set and a test set; intermediate surrogate models are constructed using the training set, and their accuracies are evaluated using the test set. By combining the errors at different sample points, an overall error estimate for each surrogate model is obtained. Two commonly used cross-validation-based error metrics are the generalized mean square cross-validation error (GMSE) and the predicted residual error sum of squares (PRESS). The formulas of the GMSE and PRESS are written as

$$GMSE = \frac{1}{N}\sum_{i=1}^{N}(y_i - \hat{y}_i)^2 \tag{3.3}$$

$$PRESS = \sum_{i=1}^{N}(y_i - \hat{y}_i)^2 \tag{3.4}$$

where y_i is the true response at x_i, N is the number of samples and $\hat{y}_i$ is the prediction value by the i-th surrogate model, which is constructed based on all $N-1$ samples other than sample x^i.

Commonly used ensemble methods based on the GMSE are discussed in the following subsections.

3.1.1.1 Inverse Proportional Averaging Method

Under the assumption that the estimated errors of different individual surrogate models are unbiased and uncorrelated, a direct way to obtain the optimal weight for each surrogate model is to calculate the ratio of its GMSE to that of the population of all individual models in an inversely proportional manner, as follows:

$$w_i = \frac{GMSE_i^{-1}}{\sum_{j=1}^{M} GMSE_j^{-1}} \tag{3.5}$$

where w_i is the weight of the i-th individual surrogate model in the combined approximation model, $GMSE_i$ is the prediction accuracy test index for the i-th surrogate model and M is the number of surrogate models.

3.1.1.2 Heuristic Computation of the Weight Coefficients

Goel et al. (2007) noted two issues related to weight selection: (1) The calculated weights should reflect the designer's confidence about the surrogate models. (2) The optimal weights should attempt to avoid adverse modelling effects, which means that the constructed model will have low generalizability. To address these two issues, Goel et al. (2007) proposed a new approach for creating an EM as an alternative to the proportional averaging method. This weighting scheme can be described as follows:

$$\begin{cases} w_i = \dfrac{w_i^*}{\sum_{j=1}^{M} w_j^*} \\ w_i^* = (E_i + \alpha \bar{E})^{\beta} \\ \bar{E} = \dfrac{1}{M}\displaystyle\sum_{i=1}^{M} E_i \\ \alpha < 1, \beta < 0 \end{cases} \tag{3.6}$$

where E_i is the error of the i-th individual surrogate model, $\bar{E}$ is the average error of all surrogate models, M is the number of surrogate models forming the EM and α and β are two constant coefficients that control the importance of the average error and the error of the individual surrogate models, respectively. A small α value and a large negative β value place greater importance on the error of the individual surrogate models, while a large α value and a small negative β value indicate high confidence in the average solution. The results of Goel et al. suggest that the EM has the best stability when $\alpha = 0.05$ and $\beta = -1$.

The weighting scheme in Eq. (3.6) is often used together with the GMSE defined in Eq. (3.3), which provides an error measure based on global data and is usually used in the form $\sqrt{GMSE_j}$ instead of E_j; consequently, this method is sometimes also called the PRESS-based weighting method.

3.1.1.3 Weight Factor Selection Based on Error Minimization

Acar et al. (2009) regard the computation of the weight coefficients for an EM as an optimization process. The weight coefficients w_i of the individual surrogate models are chosen as the design variables, and the objective function of the optimization problem is the minimization of the GMSE value for the EM. The optimization problem can be mathematically expressed as follows:

$$\begin{cases} find & w_i\ (i = 1, 2, \ldots, M) \\ \min & \varepsilon_e = Err\{\hat{y}_e(w_i, \hat{y}_i(x^k)),\ y(x^k),\ k = 1 \in N\} \\ s.t & w_i \geq 0,\ \sum_{i=1}^{M} w_i = 1 \end{cases} \tag{3.7}$$

where M is the number of surrogate models, N is the number of samples, ε_e is the global error of the EM, and $Err\{\cdot\}$ is the error metric that is utilized to measure the overall accuracy of the EM. $y(\boldsymbol{x}^k)$ is the true response for sample $\boldsymbol{x}^k$ and $\hat{y}_e(w_i, \hat{y}_i(\boldsymbol{x}^k))$ is the response predicted by the i-th individual surrogate model, which is constructed based on all samples except x^k.

If the GMSE is selected as the error metric, the global error ε_e can be determined as follows:

$$\varepsilon_e = GMSE^V = \frac{1}{N}\sum_{k=1}^{N} [\hat{y}_e(w_i, \hat{y}_i(\boldsymbol{x}_k^v)) - y(\boldsymbol{x}_k^v)]^2 \tag{3.8}$$

where N is the number of validation points, $\boldsymbol{x}_i^v$ is the i-th input variable and v is the number of points in the validation test set. Similarly, if the root mean square error (RMSE) is selected as the error metric, the weights are evaluated by minimizing the error at a set of verification points. In this case, the definition of ε_e is as follows:

$$\varepsilon_e = RMSE^V = \frac{1}{N}\sqrt{\sum_{k=1}^{N} [y(\boldsymbol{x}_k^v) - \hat{y}_e(\mathbf{w}, \boldsymbol{x}_k^v)]^2} \tag{3.9}$$

When the RMSE is used, the accuracy of the constructed EM is highly influenced by the value of N, and the optimal value of N depends on the characteristics of the problem.

The calculation process for Eq. (3.7) is generally independent of the selected global error metric. For example, other global error metrics can be used, such as the correlation coefficient between the predicted and actual responses, the multiple determination coefficient (R^2), its adjusted value (R^2_{adj}), the mean absolute error or the maximum absolute error. However, it is important to note that the accuracy of the resulting EM depends on the error metric used.

3.1.2 Optimal Weighted Ensemble of Surrogates (OWS)

Bishop (1995) used the covariance matrix of the residuals as the basis for a weighting scheme and proposed a weight coefficient calculation method called the optimal weighted ensemble of surrogates (OWS) method to combine different neural networks. In the OWS method, the EM is constructed by minimizing the mean square error (MSE) value over the design space, which can be expressed as

$$\begin{aligned} \min: & MSE_{WAS} \\ & = \frac{1}{V}\int_V e_{WAS}^2(\boldsymbol{x})d\boldsymbol{x} \\ & = \mathbf{w}^T\mathbf{C}\mathbf{w} \end{aligned} \tag{3.10}$$

Here, $e_{WAS}(\boldsymbol{x}) = y(\boldsymbol{x}) - \hat{y}_{WAS}(\boldsymbol{x})$ is the error between the true response and the response predicted by the OWS model, $\mathbf{w} = [w_1, w_2, \ldots, w_M]^T$ is the weight matrix for the M individual surrogate models, $\mathbf{C}$ is a matrix of the correlations between the prediction residuals of different surrogate models and the elements of the matrix $\mathbf{C}$ can be approximated as follows:

$$c_{ij} = \frac{1}{V}\int_V e_i(\boldsymbol{x})e_j(\boldsymbol{x})d\boldsymbol{x} \tag{3.11}$$

where $e_i(\cdot)$ and $e_j(\cdot)$ are the prediction errors associated with the i-th and j-th surrogate models, respectively.

In the actual model construction process, it is difficult to calculate the correlation matrix $\mathbf{C}$ by integrating within the domain of interest; thus, the cross-validation error vector $\tilde{\mathbf{e}}$ can be used to approximate the correlation matrix as follows:

$$c_{ij} \simeq \frac{1}{N}\tilde{\mathbf{e}}_i^T \cdot \tilde{\mathbf{e}}_j \tag{3.12}$$

where N is the number of training samples and i and j represent different surrogate models.

The optimization problem in Eq. (3.10) can be converted into the following form:

$$\begin{aligned} Min \quad & MSE_{WAS} = \mathbf{w}^T\mathbf{C}\mathbf{w} \\ s.t \quad & \mathbf{1}^T\mathbf{w} = 1 \end{aligned} \tag{3.13}$$

By solving the above-constrained optimization problem using the Lagrangian multiplier method, the weight vector can be calculated as follows:

$$\mathbf{w} = \frac{\mathbf{C}^{-1}\mathbf{1}}{\mathbf{1}^T\mathbf{C}^{-1}\mathbf{1}} \tag{3.14}$$

Matrix-based methods can reduce the computational cost of the optimization process. However, the weight coefficients obtained by Eq. (3.14) may be negative or greater than one because the approximation of the correlation matrix based on the cross-validation error magnifies the actual error value. Even by adding the constraint $0 \le w_i \le 1$ to Eq. (3.13), the poor error approximation performance of the correlation matrix still cannot be completely overcome. The most fundamental reason is that an accurate approximation of the matrix $\boldsymbol{C}$ is unrealistic in practice due to the limited design cost. In the approximate correlation matrix $\boldsymbol{C}$, the elements on the main diagonal are generally more accurate than the non-diagonal elements. Therefore, Viana et al. (2009) proposed the use of only the diagonal elements on the matrix to calculate the weight coefficients, thereby effectively ensuring that the weight coefficient values fall in the range of $[0, 1]$:

$$\mathbf{w} = \frac{\mathbf{C}_{diag}^{-1}\mathbf{1}}{\mathbf{1}^T\mathbf{C}_{diag}^{-1}\mathbf{1}} \tag{3.15}$$

3.1.3 Optimal Average Weight Method Based on Recursive Arithmetic Averaging

Since many intermediate surrogate models need to be built during the calculation of the cross-validation error, the computational cost is relatively high. When a criterion based on the smallest prediction error (RMSE or MSE) is applied, additional validation points are needed, but each surrogate model needs to be constructed only once; therefore, this approach is relatively efficient in terms of time. Zhou et al. (2011) proposed a recursive algorithm for minimizing the predicted RMSE to obtain the optimal weights. The basic framework of this method is as follows:

Input: Initial weight factors.

Step 1: Fit the training sample points $\{(x_j, y_j)\}, j = 1, 2, \ldots, T$, with N alternative models (where T is the number of training sample points).

Step 2: Calculate the predicted MSE of each candidate model at the test points,

$$e_i = \frac{1}{T}\sum_{j=1}^{T}(Sur_{ij} - \widehat{Sur_{ij}})^2, i = 1, 2, \ldots N \tag{3.16}$$

where Sur_{ij} and $\widehat{Sur_{ij}}$ are the true value and value predicted by the i-th candidate model, respectively, at the j-th test point.

Step 3: Find the worst individual surrogate model (the single surrogate model with the largest prediction error, denoted by Sur_{worst}, whose corresponding prediction MSE is denoted by $MSE_{WorstSur}$) and the best individual surrogate model (the single surrogate model with the smallest prediction error, denoted by Sur_{best}, whose corresponding prediction MSE is denoted by $MSE_{BestSur}$).

While: ($MSE_{WorstSur} - MSE_{BestSur} > tol$)

Step 4: Perform arithmetic averaging on the N candidate models to obtain the constructed EM, denoted by Sur_{ave}.

Step 5: Replace the surrogate model with the largest MSE with the EM Sur_{ave} obtained in Step 4. Thus, N new surrogate models are obtained, $N-1$ of which are unchanged. Recalculate and update the weights of the initial N individual surrogate models.

Step 6: Similar to Step 3, find the worst and best models among the newly obtained N models. If the termination condition is met, return to Step 4; otherwise, exit the loop.

End while

Output: Optimal weights.

The entire iterative process is repeated until the predicted MSE does not significantly improve, within a predetermined tolerance threshold tol (such as $tol = 0.01$).

Consider an EM consisting of N individual surrogate models $Sur_1, Sur_2 \ldots Sur_N$, each of which has a weight of w_i, where the weights satisfy $\sum_{i=1}^{N} w_i = 1$. Let the predicted response and prediction error of the i-th surrogate model Sur_i for the j-th data point be $\widehat{Sur_{ij}}$ and e_{ij}, respectively, for $j = 1, 2, \ldots, T$ (where T is the number of training points). Then, the predicted response and corresponding error of the EM for the j-th data point are $Sur_{ave}(j) = \sum_{i=1}^{N} w_i \widehat{Sur_{ij}}$ and $e_{ave}(j) = \sum_{i=1}^{N} w_i e_{ij}$, respectively. Let $W = [w_1, w_2 \ldots w_N]^T$ be the weight vector, let $E_i = [e_{i1}, e_{i2}, \ldots e_{iT}]^T$ be the prediction error vector for the i-th model Sur_i and let $e = [E_1, E_2, \ldots, E_N]$ be the prediction error matrix; then, the sum of the squared prediction errors of the simple average surrogate model Sur_{ave}, denoted by J, can be expressed as

$$J = W^T E W \tag{3.17}$$

where $E = e^T e = \begin{bmatrix} E_{11} & E_{12} & \cdots & E_{1N} \\ E_{21} & E_{22} & \cdots & E_{2N} \\ \vdots & \vdots & \vdots & \vdots \\ E_{N1} & E_{N2} & \cdots & E_{NN} \end{bmatrix}$ and $E_{ij} = E_i^T E_j = \sum_{t=1}^{N} e_{it} e_{jt}$.

Obviously, E_{ii} is the sum of the squared prediction errors of the i-th individual surrogate model Sur_i.

Compared with other methods of model combination, this method has the following characteristics:

(1) It can guarantee that the weights are within the interval $[0, 1]$ and avoid the case in which the weights are greater than 1 or less than 0, as often occurs in the OWS method.

(2) It is relatively simple and efficient when there are a large number of individual models available, since the increasing number of individual models leads to more variables in the optimization problem of solving the weights, which may slow the optimization process.

3.2 Pointwise Weighted Ensembles of Surrogate Models (EMs)

3.2.1 *Weight Coefficient Selection Based on Prediction Variance*

Zerpa et al. (2005) used the prediction variance for local error estimation and set the weight of each individual surrogate model to be inversely proportional to the pointwise prediction variance, which can be mathematically expressed as follows:

$$w_i(\boldsymbol{x}) = \frac{\frac{1}{V_i(\boldsymbol{x})}}{\sum_{j=1}^{M} \frac{1}{V_j(\boldsymbol{x})}} \tag{3.18}$$

where $V_i(\boldsymbol{x})$ is the prediction variance of the i-th surrogate model at point $\boldsymbol{x}$ and M is the number of surrogate models. Since the prediction variance is a function of $\boldsymbol{x}$, the weight of each surrogate model is also expressed as a function of $\boldsymbol{x}$. The prediction variance $V_j(\boldsymbol{x})$ of the j-th surrogate model is calculated as follows:

$$V_j(\boldsymbol{x}) = \frac{1}{N-1} \sum_{k=1}^{N} \left[y(\boldsymbol{x}^k) - y_j(\boldsymbol{x}^k)\right]^2 \tag{3.19}$$

where $\boldsymbol{x}^1, \boldsymbol{x}^2, \ldots, \boldsymbol{x}^N$ are the N samples that are closest to the prediction point $\boldsymbol{x}$.

3.2.2 *Adaptive Hybrid Function (AHF)-Based Pointwise Weighting Method*

Zhang et al. (2012) proposed an adaptive hybrid approximation model constructed based on an adaptive hybrid function (AHF). The AHF is used to formulate trust regions based on the density of available sample points and adaptively combine the features of different surrogate models. The weight of each contributing surrogate model is represented as a function of the input domain based on the local precision of that surrogate model. This approach takes advantage of each individual surrogate model to capture global and local trends in complex functional relationships. Zhang et al. (2013) also combined three types of surrogate models, namely, radial basis function (RBF), extended RBF (E-RBF) and Kriging models, with the AHF approach and investigated the influence of the sample size and the dimensionality of the design problem on the proposed method.

The construction of an EM via the AHF approach generally follows four steps:

1. Build the component surrogate models for the EM.
2. Determine the crowding-distance-based trust region, which is the boundary of the predicted output value as a function of the input vector, over the input space.
3. Characterize the local accuracy of the estimated function values (using a kernel function).
4. Form the EM based on the estimated weights.

Consider a set of training points D, which can be expressed as

$$D = \begin{pmatrix} x_1^1 & x_2^1 & \cdots & x_{n_d}^1 & y^1 \\ x_1^2 & x_2^2 & \cdots & x_{n_d}^2 & y^2 \\ \vdots & \vdots & \ddots & \vdots & \vdots \\ x_1^{n_p} & x_2^{n_p} & \cdots & x_{n_d}^{n_p} & y^{n_p} \end{pmatrix}$$

where x_j^i is the j-th component of the input vector for the i-th training point, y^i is the corresponding output, n_d is the dimensionality of the input variables and n_p is the number of training data points. The AHF-based model construction process proposed in Zhang et al. (2012) is described in detail below.

Step 1: Construction of the individual surrogate models

Different kinds of component surrogate models (such as kriging, RBF and E-RBF models) can be built based on the training set D.

Step 2: Determination of the crowding-distance-based trust region

Step 2.1: Determination of the basic model

The basic model is constructed using a smooth function with a given set of points D to address the global accuracy of the overall surrogate model. In the cited reference, the basic model was constructed using the quadratic response surface method (QRSM), and the coefficients for the QRSM model were determined using the least squares method. Of course, the basic model can also be flexibly generated using other regression methods. The mathematical formula for the QRSM model can be expressed as

$$\tilde{f}_{qrsm}(x) = a_0 + \sum_{i=1}^{n_d} b_i x_i + \sum_{i=1}^{n_d} c_{ii} x_i + 2 \sum_{i=1}^{n_d-1} \sum_{j>i}^{n_d} c_{ij} x_i x_j \tag{3.20}$$

where x_i are the input parameters and the variables a_0, b_i and c_{ij} are the unknown coefficients determined via the least squares method.

Step 2.2: Generation of the trust region boundaries

In this step, a trust region is constructed based on the density of the sample points; more specifically, this region is called the crowding-distance-based trust region (CD-TR). A set of points is generated from the basic model, and the crowding distance is evaluated for each point. The trust region boundaries can be adaptively calculated based on the obtained crowding distances at the training points and the predictions of the basic model. The basic model can be relaxed along either output axis to obtain the trust region boundaries for the surrogate model.

The crowding distance is utilized to estimate the density of the training points around any point in the basic model. A larger value of the crowding distance indicates a lower sample density (i.e. there are fewer training points around that point), and the accuracy of the surrogate model is expected to lie within the boundary. Therefore, a relatively broad boundary is needed at a low-density point. Based on the crowding distance values obtained at different points in the basic model, the adaptive boundaries of the CD-TR are constructed. The crowding distance at the i-th point (CD^i) in the basic model is calculated as follows:

$$CD^i = \sum_{j=1}^{n_p} \left\| x^j - x^i \right\|^2 \tag{3.21}$$

where n_p is the number of training points. A parameter ρ^i is used to reflect the local density at the i-th training point; this parameter is expressed as follows:

$$\rho^i = \frac{1}{CD^i} \tag{3.22}$$

The parameters ρ^i can be normalized to α^i:

$$\alpha_i = \frac{\max(\rho) - \rho^i}{\max(\rho) - \min(\rho)} \tag{3.23}$$

The adaptive distance d_i between the i-th point in the basic model and the corresponding point on the boundary can be written as

$$d^i = (1 + \alpha_i) \times \max_{j \in D} \left| \tilde{f}_{qrsm}(x^j) - y^j \right| \tag{3.24}$$

where D is the training data set. In Eq. (3.25), j denotes a training point and i denotes a set of uniformly distributed points selected from the basic model. Therefore, the formula for the adaptive distance can be divided into two parts as given below:

(1) $\max_{j \in D} \left| \tilde{f}_{qrsm}(x^j) - y^j \right|$ is a constant, which is used to ensure that all training points are located within the boundaries.

(2) $\alpha_i \times \max_{j \in D} \left| \tilde{f}_{qrsm}(x^j) - y^j \right|$ is the adaptive distance, whose value changes with different distance coefficients α_i.

The crowding distance is evaluated for each selected point based on a previously formulated α. The range of the boundary region is normalized with respect to the maximum deviation of the training data from the basic model. Here, $\tilde{f}_{qrsm}(x^j)$ is the output value predicted by the QRSM at the j-th training point and $\left| \tilde{f}_{qrsm}(x^j) - y^j \right|$ is the distance between the prediction for the j-th training point and the basic model. Thus, two sets of points, D^U and D^L, are obtained to construct the two boundaries, as follows:

$$D^U = \begin{pmatrix} x_1^1 & x_2^1 & \cdots & x_{n_d}^1 & y^1 + d^1 \\ x_1^2 & x_2^2 & \cdots & x_{n_d}^2 & y^2 + d^2 \\ \vdots & \vdots & \ddots & \vdots & \vdots \\ x_1^{n_p} & x_2^{n_p} & \cdots & x_{n_d}^{n_p} & y^{n_p} + d^{n_p} \end{pmatrix}$$

$$D^L = \begin{pmatrix} x_1^1 & x_2^1 & \cdots & x_{n_d}^1 & y^1 - d^1 \\ x_1^2 & x_2^2 & \cdots & x_{n_d}^2 & y^2 - d^2 \\ \vdots & \vdots & \ddots & \vdots & \vdots \\ x_1^{n_p} & x_2^{n_p} & \cdots & x_{n_d}^{n_p} & y^{n_p} - d^{n_p} \end{pmatrix}$$

The upper and lower boundaries are estimated via the QRSM using the generated data points D^U and D^L, respectively. The boundaries of the trust region can be constructed using the QRSM and the two data sets D^U and D^L, and the probability associated with each individual surrogate model can be calculated based on the trust region boundaries.

Step 3: Estimation of measurement accuracy

To represent the uncertainty of the predicted responses, a metric called the accuracy measure of surrogate modelling is adopted. A kernel function, which is written as a function of the output parameters, is used to model the uncertainty of the predicted function values at a certain point in the input variable domain. The coefficients in the kernel function can be expressed as functions of the input variables, characterizing the accuracy measure of surrogate modelling over the entire input domain.

The kernel function used to calculate the accuracy measure of surrogate modelling must have the following properties:

(1) The maximum value of the kernel function must be one, which corresponds to the actual output.
(2) The value of the kernel function must be equal to the small specified tolerance at the upper and lower boundaries of the trust region.

(3) The function must increase monotonically from any boundary to the actual output value.
(4) The function must be continuous.

Zhang et al. (2013) suggested the following kernel function for use in calculating the accuracy metric:

$$P(z) = a\exp\left[-\frac{(z-u)^2}{2\sigma^2}\right] \tag{3.25}$$

where the value of the amplitude coefficient a is specified as 1 and u and σ represent the mean value and standard deviation, respectively, of the kernel function. Other kernel functions with similar properties can also be used for this purpose.

The distances between the two boundaries and the output values $f_L^i(x^i)$ and $f_U^i(x^i)$ at each training data point x^i are normalized. Thus, the maximum value of the estimated accuracy metric (kernel function) is 1, corresponding to the actual output value $y(x^i)$, and the minimum value is 0.1, corresponding to the boundaries (within the trust region). With this convention, the true response at a particular training point does not necessarily lie in the middle of the two boundaries. To ensure that the kernel function is continuous, it is divided into two parts with the same mean but different standard deviations. The kernel function can then be represented as follows:

$$P(x^i) = \begin{cases} a\exp\left\{-\frac{[y(x^i)-\mu(x^i)]^2}{2\sigma_1^2(x^i)}\right\} & 0 \le y(X^i) < \mu(X^i) \\ a\exp\left\{-\frac{[y(x^i)-\mu(x^i)]^2}{2\sigma_2^2(x^i)}\right\} & \mu(X^i) \le y(X^i) \le 1 \end{cases} \tag{3.26}$$

The parameters σ_1 and σ_2 are controlled with respect to one-tenth of the required maximum width (Δx_{10}) and are calculated as follows:

$$\sigma_1(x^i) = \frac{\Delta z_{10}(x^i)}{2\sqrt{2\,In 10}} = \frac{2[\mu(x^i) - f_L^i(x^i)]}{2\sqrt{2\,In 10}} = \frac{\mu(x^i)}{\sqrt{2\,In 10}} \tag{3.27}$$

$$\sigma_2(x^i) = \frac{\Delta z_{10}(x^i)}{2\sqrt{2\,In 10}} = \frac{2[f_U^i(x^i) - \mu(x^i)]}{2\sqrt{2\,In 10}} = \frac{1-\mu(x^i)}{\sqrt{2\,In 10}} \tag{3.28}$$

where

$$P(\mu \pm 0.5\Delta z_{10}) = \frac{1}{10} \tag{3.29}$$

By using Eq. (3.26), the precision coefficient $\mu(x^i)$ can be calculated for the i-th training point and the coefficient μ is expressed in terms of a polynomial response surface related to the input variable x_j^i.

Step 4: Combination of individual surrogate models

The AHF-based surrogate model is constructed by adaptively calculating the weights of three component surrogate models (RBF, E-RBF and kriging models). The AHF model is the weighted sum of the component surrogate models as follows:

$$\tilde{f}_{AHF} = \sum_{i=1}^{ns} w_i \tilde{f}_i(x) \tag{3.30}$$

Here, ns is the number of component surrogate models, $\tilde{f}_i(x)$ represents the prediction response of the i-th surrogate model and the weights w_i are calculated in accordance with the estimated measure of accuracy as follows:

$$w_i(x) = \frac{P_i(x)}{\sum_{i=1}^{ns} P_i(x)} \tag{3.31}$$

where $P_i(x)$ is the estimated measure of accuracy for the i-th surrogate model at point x.

3.2.3 *Pointwise Weighted EM Using* v *Nearest Points Cross-Validation*

Lee and Choi (2014) proposed a method for determining the weights at unknown points by using the cross-validation errors at v known sample points. The EM is constructed using the weighted sum formula of one of the other methods described above. The weight coefficients can be calculated as follows:

$$w_i(\boldsymbol{x}) = \frac{\frac{1}{vCV_i(\boldsymbol{x})}}{\sum_{j=1}^{M} \frac{1}{vCV_j(\boldsymbol{x})}} \tag{3.32}$$

where

$$vCV_i(\boldsymbol{x}) = \sqrt{\frac{1}{v}\sum_{k=1}^{v}\left\{\hat{y}_i(\boldsymbol{x}) - \hat{y}_i^{(-k)}(\boldsymbol{x})\right\}^2} \tag{3.33}$$

The error defined in Eq. (3.33) is called the v-nearest-point cross-validation error (vCV). From the formula for vCV, it can be seen that the calculation of vCV

requires the surrogate model to be reconstructed only v times, while the calculation of the GMSE requires *nexp* reconstructions. Thus, the computational cost of vCV is lower than that of the GMSE, and this efficiency improvement will become more obvious as *nexp* increases. In addition, vCV is a local error and is expressed as a function of $\boldsymbol{x}$, while the GMSE is an average error and yields a weight for each component surrogate model that remains the same over the entire design space.

In interpolation methods, such as the Kriging and RBF methods, surrogate models are constructed by interpolating from the sample points; thus, the predicted responses are equal to the true responses at the training points. For models constructed using regression methods, such as Poisson regression (PR) and support vector regression (SVR), there are generally some discrepancies between the predicted and true responses at the training points. If both interpolation and regression models are used as the component models in Eq. (3.32), the resulting EM will tend not to yield very accurate predictions due to the influence of the regression model(s). To more reasonably integrate predictions from interpolation models and regression models, a modified v-nearest-point cross-validation error (vCV) can be used as given below:

$$vCV_i^*(\boldsymbol{x}) = \begin{cases} \alpha(\boldsymbol{x})\cdot\sqrt{\dfrac{1}{v}\displaystyle\sum_{k=1}^{v}\left\{\hat{y}_i(\boldsymbol{x}) - \hat{y}_i^{(-k)}(\boldsymbol{x})\right\}^2}, & \text{for interpolation models} \\ \sqrt{\dfrac{1}{v}\displaystyle\sum_{k=1}^{v}\left\{\hat{y}_i(\boldsymbol{x}) - \hat{y}_i^{(-k)}(\boldsymbol{x})\right\}^2}, & \text{for regression models} \end{cases} \tag{3.34}$$

For interpolation models, a control function $\alpha(x)$ is introduced.

To create an interpolated EM, the control function $\alpha(x)$ must satisfy the following requirements:

$$\lim_{d_1(\boldsymbol{x})\to 0}\alpha(\boldsymbol{x}) = 0, \qquad \lim_{d_1(\boldsymbol{x})\to d_2(\boldsymbol{x})}\alpha(\boldsymbol{x}) = 1, \tag{3.35}$$

where $d_1(\boldsymbol{x})$ is the distance between $\boldsymbol{x}$ and the training point that is the closest to $\boldsymbol{x}$ and $d_2(\boldsymbol{x})$ is the distance between $\boldsymbol{x}$ and the second closest training point to $\boldsymbol{x}$. Lee and Choi (2014) suggested using the following third-order polynomial as the control function:

$$\alpha(\boldsymbol{x}) = 3\left(\frac{d_1(\boldsymbol{x})}{d_2(\boldsymbol{x})}\right)^2 - 2\left(\frac{d_1(\boldsymbol{x})}{d_2(\boldsymbol{x})}\right)^3 \tag{3.36}$$

The pointwise weights of each component model that are obtained using the improved v-nearest-point cross-validation error (vCV^*) can be calculated as follows:

$$w_i(\boldsymbol{x}) = \frac{\frac{1}{vCV_i^*(\boldsymbol{x})}}{\sum_{j=1}^{M} \frac{1}{vCV_j^*(\boldsymbol{x})}} \tag{3.37}$$

3.2.4 *Optimal Weighted Pointwise Ensemble (OWPE) Method*

To avoid the selection of basis functions for RBF models and generate improved predictions, Liu et al. (2016) proposed the optimal weighted pointwise ensemble (OWPE) method for the construction of an EM. Compared with other existing ensemble methods, the main differences of the OWPE method are as follows:

(1) For the weights at the sample points, the 0–1 weight strategy is utilized, which can emphasize the local accurate prediction accuracy of stand-alone RBF model.
(2) For the weights at unobserver points, pointwise weights are calculated by solving an optimization problem minimizing GMSE, which can adopt the characteristics of stand-alone RBF models.

The flowchart of the OWPE method is shown in Fig. 3.1, and the details of the procedures are given below.

Step 1: Construct k different RBF models

Using the same sample set, build k different RBF models with different basis functions.

Step 2: Determine the weights at the observed points

Suppose that for a given function f, k RBF models can be constructed with different basis functions. If cross-validation errors are used to represent the local errors of the RBF models, the error of the j-th RBF model $\hat{f}_j$ at sample point $\boldsymbol{x}$ can be expressed as

$$\bar{e}_i^j = Ge_i^j, i = 1, 2, \ldots m; j = 1, 2, \ldots, k \tag{3.38}$$

where $G_j = \frac{GMSE_j}{\max\limits_{i=1,2,\ldots,k}\{GMSE_i\}}$, which represents the normalized global error of $\hat{f}_j$. A small value of G_j indicates that the RBF model has good global accuracy. $\bar{e}_i^j$ is the cross-validation error of $\hat{f}_j$ at point $\boldsymbol{x}_i$. A small value of $\bar{e}_i^j$ indicates that the RBF model has good local performance.

If the $\bar{e}_i^j$ value of $\hat{f}_j$ is the smallest among all component RBF models, then $\hat{f}_j$ is considered to have the highest prediction accuracy near point $\boldsymbol{x}_i$. To take full advantage of the high-precision predictions of this model in this region, a 0–1 weighting strategy is used, i.e. the weight w_{ij} of $\hat{f}_j$ at point $\boldsymbol{x}_i$ is set to 1 and the weights of the remaining component RBF models at point $\boldsymbol{x}_i$ are set to zero.

Fig. 3.1 Flowchart of the OWPE method

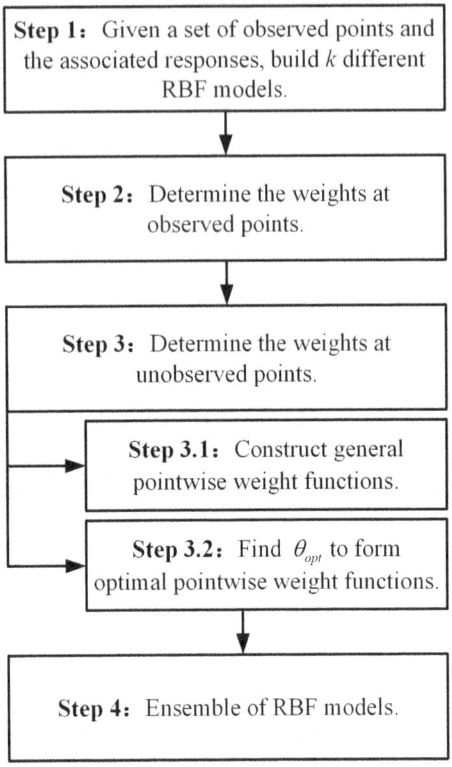

The above procedure is repeated to obtain a set of weights $W_j = \{w_{1j}, w_{2j}, \ldots, w_{mj}\}^T$ for each $\hat{f}_j$ at all sample points, all elements of which are either 0 or 1. These weights are called observation weights.

Step 3: Determine the weights at unobserved points

Step 3.1: Construct general pointwise weight functions.

The weights at the observed sample points obtained in the last step can be used to predict the weights at unobserved points. For the j-th RBF model $\hat{f}_j$, its weight $w_j(\boldsymbol{x})$ at any unobserved point x can be expressed using a weighted formula of inverse ratios of distance:

$$\omega_j(\boldsymbol{x}) = \sum_{i=1}^{m} \omega_j^i(\boldsymbol{x}) = \sum_{i=1}^{m} \frac{d_i^{-\theta_i} W_{ij}}{\sum d_i^{-\theta_i}}, \boldsymbol{x} \neq \boldsymbol{x}_i, j = 1, 2, \ldots, k \qquad (3.39)$$

where $\omega_j^i(\boldsymbol{x})$ is the i-th element of $\omega_j(\boldsymbol{x})$ and reflects how much the observation weight W_{ij} contributes to the weight $\omega_j(\boldsymbol{x})$ at an unobserved point. A larger value of $\omega_j^i(\boldsymbol{x})$ indicates that $\omega_j(\boldsymbol{x})$ is closer to W_{ij}. $d_i = \|\boldsymbol{x} - \boldsymbol{x}_i\|$ is the Euclidean distance between the observed point $\boldsymbol{x}_i$ and the unobserved point $\boldsymbol{x}$ and the index θ_i is a coefficient that influences the attenuation rate of w_{ij}. The value of $\omega_j^i(\boldsymbol{x})$ decreases with increasing d_i, and a larger value of θ_i leads to a slower decrease. Equation (3.39) shows that the weight $\omega_j(\boldsymbol{x})$ depends on the weights at all sample points.

If $W_{ij} = 1$, this means that $\hat{f}_j$ has the highest prediction accuracy near the point $\boldsymbol{x}_i$. In this case, $\omega_j^i(\boldsymbol{x}) = \frac{d_i^{-\theta_i}}{\sum d_i^{-\theta_i}}$ in Eq. (3.39). Conversely, if $W_{ij} = 0$, then $\omega_j^i(\boldsymbol{x}) = 0$, and the weight at this sample point has no effect on $\omega_j(\boldsymbol{x})$. Therefore, an observation weight of 1 can be regarded as an active observation weight; conversely, a weight of 0 is called a frozen observation weight. Due to the 0–1 weighting strategy, one and only one model among the component RBF models will provide an active observation weight at point $\boldsymbol{x}_i$.

Step 3.2: Find θ_{opt} to form optimal pointwise weight functions.

The parameter θ_i describes the attenuation rate of the active observation weight W_{ij}; in other words, the value of θ_i determines the domain of influence of observation weight W_{ij}. Since the active observation weights at the m known points may belong to different RBF models, the value of the parameter θ_i should be determined individually for each corresponding RBF model and thus can be defined as follows:

$$\theta_i = B_i\theta \tag{3.40}$$

where B_i is the value of the normalized global accuracy metric of the component RBF model that has an active observation weight. If $\bar{f}_u$ is the active RBF model, then $B_i = \frac{1/GMSE_u}{\max\limits_{j=1,2,\ldots,k}\{1/GMSE_j\}}$. θ is a constant attenuation factor in the EM and is scaled as shown above for different component RBF models.

Finally, the pointwise weight function in the design space can be expressed as follows:

$$\omega_j(x) = \begin{cases} \sum\limits_{i=1}^{m} \dfrac{d_i^{-B_i\theta} W_{ij}}{\sum d_i^{-B_i\theta}} & (\boldsymbol{x} \neq \boldsymbol{x}_i) \\ W_{ij} & (\boldsymbol{x} = \boldsymbol{x}_i) \end{cases} ,j = 1, 2, \ldots, k \tag{3.41}$$

The attenuation constant coefficient θ has a great influence on the point-by-point weight function and further affects the prediction accuracy of the ensemble of RBF models. A smaller θ value results in smaller ranges of influence of the observation weights. In the extreme case of $\theta = 0$, the pointwise ensemble degenerates to an average ensemble. For realistic problems, the selection of the most appropriate θ value depends on the characteristics of the problem of interest. In the OWPE method, the optimal attenuation constant θ_{opt} is obtained by solving an optimization problem that minimizes the GMSE of the EM:

$$\begin{aligned}&\textit{Find } \theta_{opt}\\&\textit{to } \min GMSE_e = \sqrt{\frac{1}{m}\sum_{i=1}^{m}\left(f(\boldsymbol{x}_i) - \hat{f}_e^{(-i)}(\boldsymbol{x}_i, \theta)\right)^2}\\&\textit{where } \hat{f}_e^{(-i)}(\boldsymbol{x}_i, \theta) = \sum_{j=1}^{k}\omega_j^{(-i)}(\boldsymbol{x}_i, \theta)\hat{f}_e^{(-i)}(\boldsymbol{x}_i)\end{aligned} \tag{3.42}$$

where $\hat{f}_e^{(-i)}(\boldsymbol{x}_i, \theta)$ is the predicted value at point $\boldsymbol{x}_i$ that is obtained from the ensemble of RBF models that is constructed based on all sample points except $\boldsymbol{x}_i$, whereas $\omega_j^{(-i)}(\boldsymbol{x}_i, \theta)$ is the predicted value at point $\boldsymbol{x}_i$ that is obtained using the weight equation that is constructed based on all existing points except $\boldsymbol{x}_i$. Once θ_{opt} has been obtained, it can be substituted into Eq. (3.41).

Step 4: Combination of the RBF models

After solving for the optimal attenuation coefficient and calculating the optimal pointwise weight for each component RBF model, the ensemble model can be written as follows:

$$\hat{f}_e(\boldsymbol{x}) = \sum_{j=1}^{k} W_j(\boldsymbol{x})\hat{f}_j(\boldsymbol{x}) \tag{3.43}$$

Two main tasks contribute to the modelling cost: calculating the GMSE metric and solving the optimization problem. Each component RBF model needs to be constructed m times to calculate the GMSE metric; consequently, the computational cost may increase exponentially as the sample set becomes larger. To relieve the computational burden, the calculation process can be implemented in a parallel way. For the observation weight matrix $\omega_j^{(-i)}(\boldsymbol{x}_i, \theta)$ in the optimization problem, Liu et al. (2016) suggested reusing the observation weight matrix W in Step A by simply deleting its i-th row. By adopting the above two methods, a near-optimal EM can be obtained.

References

Acar E, Rais-Rohani M (2009) Ensemble of metamodels with optimized weight factors. Struct Multidiscip Optim 37:279–294

Bishop CM (1995) Neural networks for pattern recognition. Oxford university press

Goel T, Haftka RT, Shyy W, Queipo NV (2007) Ensemble of surrogates. Struct Multidiscip Optim 33:199–216

Lee Y, Choi D-H (2014) Pointwise ensemble of meta-models using v nearest points cross-validation. Struct Multidiscip Optim 50:383–394

Liu H, Xu S, Wang X, Meng J, Yang S (2016) Optimal weighted pointwise ensemble of radial basis functions with different basis functions. AIAA J 3117–3133

Viana FA, Haftka RT, Steffen V (2009) Multiple surrogates: how cross-validation errors can help us to obtain the best predictor. Struct Multidiscip Optim 39:439–457

Zerpa LE, Queipo NV, Pintos S, Salager J-L (2005) An optimization methodology of alkaline–surfactant–polymer flooding processes using field scale numerical simulation and multiple surrogates. JoPS Eng 47:197–208

Zhang J, Chowdhury S, Messac A (2012) An adaptive hybrid surrogate model. Struct Multidiscip Optim 46:223–238

Zhang J, Chowdhury S, Messac A, Castillo L (2013) Adaptive hybrid surrogate modeling for complex systems. AIAA J 51:643–656

Zhou XJ, Ma YZ, Li XF (2011) Ensemble of surrogates with recursive arithmetic average. 44:651–671

Chapter 4
Multi-fidelity Surrogate Models

Simulation models are treated as black boxes that generate input–output correspondences. Nevertheless, designers need to choose simulation models with appropriate fidelities to obtain qualities of interest (QOIs) at affordable cost levels. Generally, high-fidelity (HF) simulation models can provide more reliable and accurate simulation results than low-fidelity (LF) models. Consider the example of designing an airfoil, which is an aerodynamic component, the available simulation models for obtaining aerodynamic coefficients may differ in terms of their resolutions (e.g. coarse meshes versus refined meshes in finite element models), levels of abstraction (e.g. two-dimensional models versus three-dimensional models) or mathematical descriptions (e.g. the Euler non-cohesive equations versus the Navier–Stokes viscous Newton equations). Relying entirely on HF models to obtain QOIs for constructing a surrogate model is always time-consuming and may even be computationally prohibitive. On the other hand, LF models are considerably less computationally demanding. However, the QOIs obtained from LF models may result in inaccurate surrogate models or even distorted ones.

A promising way to achieve a trade-off between prediction accuracy and computational cost in modelling is to integrate the information from both HF and LF simulations by constructing a multi-fidelity (MF) surrogate model (Viana et al. 2014; Chen et al. 2016; Zhou et al. 2017). MF modelling is based on the assumption that in addition to an HF model that is sufficiently accurate but has a high computational cost, an LF model is used that is less accurate but also considerably less computationally demanding (Viana et al. 2014). The three main approaches for constructing MF models are scaling-function-based MF modelling, space mapping and co. These approaches are summarized as follows:

(1) Scaling-function-based MF modelling

Scaling-function-based MF modelling approaches can be divided into three distinct types. First, in the multiplicative scaling approach, a scaling function is constructed to represent the ratio between the HF and LF models (Burgee et al. 1996; Liu and Collette 2014). Second, in the additive scaling approach, a scaling function is

P. Jiang et al., *Surrogate Model-Based Engineering Design and Optimization*,
Springer Tracts in Mechanical Engineering,
https://doi.org/10.1007/978-981-15-0731-1_4

constructed to capture the differences between the HF and LF models (Viana et al. 2009; Sun et al. 2012; Zhou et al. 2015, 2016b). Finally, in the hybrid scaling approach, scaling functions are constructed to utilize the advantages of both the multiplicative and additive scaling approaches (Gano et al. 2005; Zheng et al. 2013; Tyan et al. 2015). It is important to note that the multiplicative scaling approach may become invalid when the values of the LF model are equal to zero at some sample points. This property, to some extent, limits the ability to use this approach to solve constrained design optimization problems because finding an optimum generally requires some constraints to be active.

(2) Space mapping (SM)

The key idea behind SM approaches is to construct an MF surrogate model by mapping the HF model's parameter space to the LF model's parameter space or mapping the output space of the LF model to that of the HF model. An obvious advantage of SM is that it allows the design parameter vectors of the LF and HF models to have different dimensions (Bandler et al. 2004; Rayas-Sanchez 2016).

(3) Co-kriging

Co-kriging was originally developed in the geostatistics community (Journel and Huijbregts 1978) and was then extended to the integration of computational models with different fidelity levels by Kennedy and O'Hagan (2000). In Kennedy and O'Hagan's approach (Journel and Huijbregts 1978), a Gaussian process (GP) model was adopted to approximate the HF response as the sum of a scaled LF response and a discrepancy function. Many variations of co-kriging methods are gaining popularity today, such as the incorporation of gradient information in co-kriging (Journel and Huijbregts 1978), the design of a nested sampling approach for co-kriging (Xiong et al. 2013), the combination of co-kriging methods with expected improvement (EI) functions in sequential optimization design (Huang et al. 2006; Forrester et al. 2007) and parameter simplification during the solution process for co-kriging (Zimmermann and Han 2010; Han et al. 2012; Le Gratiet and Garnier 2014; Hu et al. 2017).

In the remainder of this chapter, a detailed introduction to these three MF modelling approaches will be provided.

4.1 Scaling-Function-Based Approaches

4.1.1 Multiplicative Scaling Approach

The multiplicative scaling approach was first proposed by Haftka (Haftka 1991; Gano et al. 2005). In this approach, a scaling factor is adopted to represent the ratio between the HF and LF models at the HF sample points. Then, a scaling function is constructed to fit the relationships between the design variables and the corresponding output. The MF surrogate model $\hat{f}_{MF}(\boldsymbol{x})$ can be obtained using the following equation:

$$\hat{f}_{MF}(\boldsymbol{x}) = \hat{f}_l(\boldsymbol{x}) \cdot \hat{\beta}(\boldsymbol{x}) \tag{4.1}$$

where $\hat{f}_l(\boldsymbol{x})$ is the LF model and $\hat{\beta}(\boldsymbol{x})$ is the scaling function.

Based on the LF sample points $\boldsymbol{x}_l = \left\{\boldsymbol{x}_{l,1}, \boldsymbol{x}_{l,2}, \ldots, \boldsymbol{x}_{l,m_l}\right\}$ and the corresponding responses $\boldsymbol{f}_l = \left\{f_{l,1}, f_{l,2}, \ldots, f_{l,m_l}\right\}$, the LF surrogate model $\hat{\boldsymbol{f}}_l(\boldsymbol{x})$ can be constructed (Ollar et al. 2017). Then, for the given HF sample points $\boldsymbol{x}_h = \left\{\boldsymbol{x}_{h,1}, \boldsymbol{x}_{h,2}, \ldots, \boldsymbol{x}_{h,m_h}\right\}$ and the corresponding responses $\boldsymbol{f}_h = \{f_{h,1}, f_{h,2}, \ldots, f_{h,m_h}\}$, the scaling factors $\beta(\boldsymbol{x}_h) = \left\{\beta(\boldsymbol{x}_{h,1}), \beta(\boldsymbol{x}_{h,2}), \ldots, \beta(\boldsymbol{x}_{h,m_h})\right\}$ between the HF and LF models at the locations $\boldsymbol{x}_{h,i}$ can be calculated as follows:

$$\beta(\boldsymbol{x}_{h,i}) = \frac{f^h(\boldsymbol{x}_{h,i})}{\hat{f}^l(\boldsymbol{x}_{h,i})} \tag{4.2}$$

Based on $\boldsymbol{x}_h$ and $\beta(\boldsymbol{x}_h)$, the scaling function $\hat{\beta}(\boldsymbol{x}_h)$ can be constructed using a modelling technique.

A one-dimensional (1D) numerical function will be used here to demonstrate the process of constructing an MF surrogate model with multiplicative scaling. The mathematical expressions for the 1D function are given in Eq. (4.3), and the function is visualized in Fig. 4.1.

$$\begin{aligned} y_h &= 0.5\sin(4\pi\sin(x+0.5)) + \frac{(x+0.5)^2}{3} \\ y_l &= 0.5\sin(4\pi\sin(1.1x+0.4)) + \frac{(1.1x+0.5)^2}{3} - 0.2 \\ x &\in [0, 1]. \end{aligned} \tag{4.3}$$

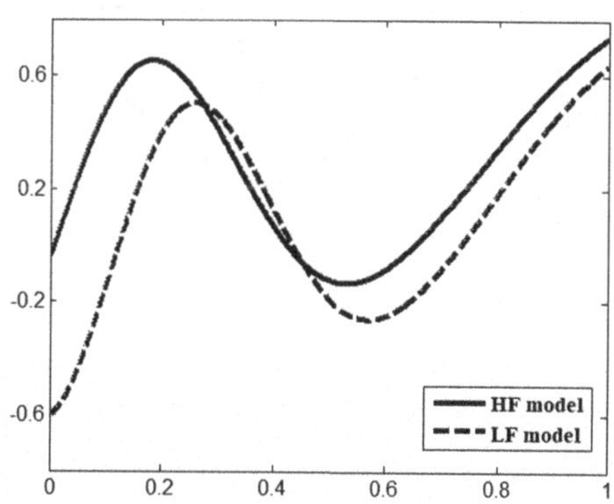

Fig. 4.1 HF and LF models of the 1D numerical function

Eleven LF sample points $\boldsymbol{x}_l = \{0, 0.1, 0.2, 0.3, 0.4, 0.5, 0.6, 0.7, 0.8, 0.9, 1.0\}$ and six HF sample points $\boldsymbol{x}_h = \{0, 0.2, 0.4, 0.6, 0.8, 1.0\}$ are selected for constructing the MF surrogate model. First, the LF model is evaluated at the points in $\boldsymbol{x}_l$ to obtain the LF responses; then, the LF surrogate model can be constructed using a modelling technique to obtain, e.g. a kriging model. Subsequently, the scaling factors can be obtained using Eq. (4.2), and the scaling function $\hat{\beta}(\boldsymbol{x}_h)$ can be constructed. The LF surrogate model and the multiplicative scaling function are visualized in Fig. 4.2a. Based on the LF surrogate model and the scaling function, the final MF surrogate model can be obtained in accordance with Eq. (4.1), as shown in Fig. 4.2b.

4.1.2 Additive Scaling Approach

Lewis et al. (Lewis and Nash 2000) developed the additive scaling approach for MF modelling. In the additive scaling approach, the scaling factors are defined as the differences between the HF and LF models at the HF sample points. Once the scaling factors are obtained, they are fitted using the scaling function to map the differences between the HF and LF models. The MF surrogate model $\hat{\boldsymbol{f}}_{MF}(\boldsymbol{x})$ can be expressed as

$$\hat{\boldsymbol{f}}_{MF}(\boldsymbol{x}) = \hat{\boldsymbol{f}}_l(\boldsymbol{x}) + \hat{\boldsymbol{C}}(\boldsymbol{x}) \tag{4.4}$$

Here, $\hat{\boldsymbol{f}}_l(\boldsymbol{x})$ represents the LF model and is constructed using a modelling technique based on the LF sample points $\boldsymbol{x}_l = \left\{\boldsymbol{x}_{l,1}, \boldsymbol{x}_{l,2}, \ldots, \boldsymbol{x}_{l,m_l}\right\}$ and the corresponding responses $\boldsymbol{f}_l = \left\{f_{l,1}, f_{l,2}, \ldots, f_{l,m_l}\right\}$, and $\hat{\boldsymbol{C}}(\boldsymbol{x})$ is the scaling function.

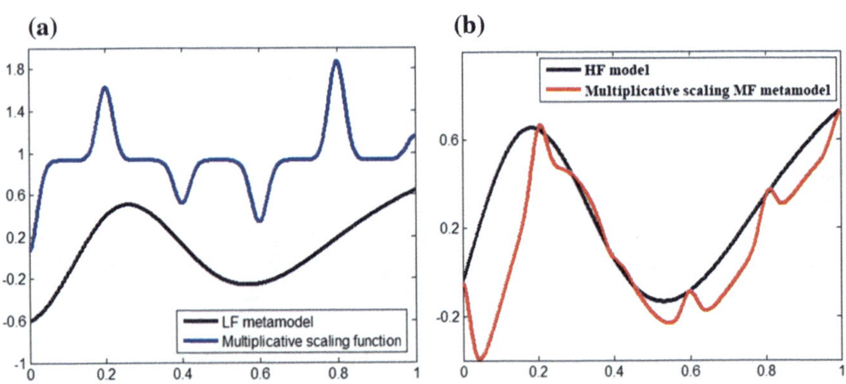

Fig. 4.2 a The LF surrogate model and the multiplicative scaling function and b the MF surrogate model with multiplicative scaling for the 1D function

Based on the HF sample points $\boldsymbol{x}_h = \left\{\boldsymbol{x}_{h,1}, \boldsymbol{x}_{h,2}, \ldots, \boldsymbol{x}_{h,m_h}\right\}$ and the corresponding responses $\boldsymbol{f}_h = \left\{f_{h,1}, f_{h,2}, \ldots, f_{h,m_h}\right\}$, the differences $\boldsymbol{C}(\boldsymbol{x}_h) = \left\{c(\boldsymbol{x}_{h,1}), c(\boldsymbol{x}_{h,2}), \ldots, c(\boldsymbol{x}_{h,m_h})\right\}$ between the HF and LF surrogate models at the locations $\boldsymbol{x}_{h,i}$ can be calculated as

$$\boldsymbol{C}(\boldsymbol{x}_{h,i}) = f_h(\boldsymbol{x}_{h,i}) - \hat{f}_l(\boldsymbol{x}_{h,i}) \tag{4.5}$$

Then, the scaling function can be approximated using a modelling technique based on $\boldsymbol{x}_h$ and $\boldsymbol{C}(\boldsymbol{x}_h)$. The 1D numerical function expressed in Eq. (4.3) will again be used to demonstrate the process of constructing an MF surrogate model with additive scaling. Based on the LF sample points $\boldsymbol{x}_l$ and the corresponding responses, an LF surrogate model can be constructed, e.g. a kriging model. The differences $\boldsymbol{C}(\boldsymbol{x}_h)$ can then be obtained in accordance with Eq. (4.5), and the scaling function $\hat{\boldsymbol{C}}(\boldsymbol{x})$ can be constructed. The LF surrogate model and the scaling function $\hat{\boldsymbol{C}}(\boldsymbol{x})$ are visualized in Fig. 4.2a. Based on the LF surrogate model and the scaling function, the final MF surrogate model can be obtained using Eq. (4.4), as shown in Fig. 4.3b.

4.1.3 Hybrid Scaling Approach

Gano et al. (2005) proposed the hybrid scaling approach to make use of the advantages of both the multiplicative scaling approach and the additive scaling approach. The variable-fidelity surrogate model $\hat{f}_{MF}(\boldsymbol{x})$ can be obtained using the following equation:

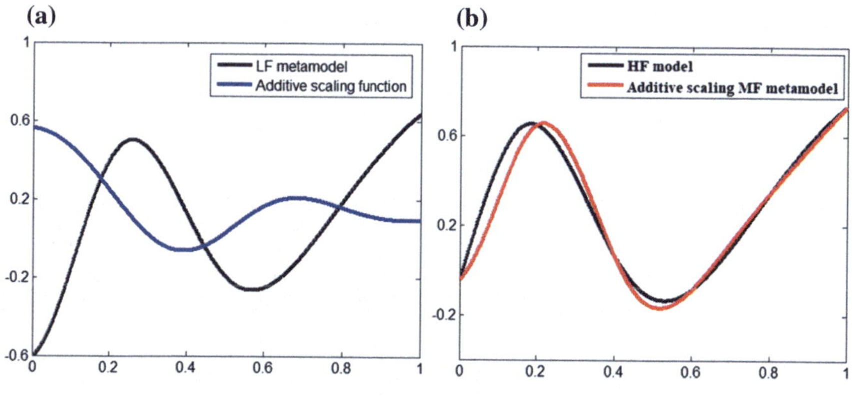

Fig. 4.3 **a** The LF surrogate model and the additive scaling function and **b** the MF surrogate model with additive scaling for the 1D function

$$\hat{f}_{MF}(\boldsymbol{x}) = \omega(\hat{f}^l(x)\hat{\beta}(\boldsymbol{x})) + (1-\omega)(\hat{f}^l(x) + \hat{C}(\boldsymbol{x})) \tag{4.6}$$

where $\hat{f}^l(x)\hat{\beta}(\boldsymbol{x})$ and $\hat{f}^l(x) + \hat{C}(\boldsymbol{x})$ are the MF surrogate models with multiplicative scaling and additive scaling, respectively and ω denotes the weighting factor reflecting the ratio between the contributions of these two different scaling approaches. The value of ω is often selected based on the knowledge and experience of the designers (Van Nguyen et al. 2015).

MF surrogate models with multiplicative scaling and additive scaling were constructed for the same 1D numerical function in Sects. 4.1.1 and 4.1.2, respectively. The corresponding MF surrogate model with hybrid scaling that is obtained by setting $\omega = 0.5$ is shown in Fig. 4.4.

4.1.4 Examples and Results

Four well-known numerical problems with different characteristics with regard to variable-fidelity modelling (Huang et al. 2006; Zheng et al. 2014; Cheng et al. 2015; Zhou et al. 2016a) are utilized here to test the performance of the three scaling-function-based MF modelling approaches introduced above. The expressions for the corresponding mathematical functions are given as follows. In these problems, y_h denotes the HF model that needs to be approximated and y_l denotes the LF model. The features of the four numerical examples are listed in Table 4.1.

Problem 1 (P1)

$$\begin{aligned} y_l &= ((0.5x_1)^2 + 0.8x_2 - 11)^2 + ((0.8x_2)^2 + 0.5x_1 - 7)^2 + x_2^3 - (x_1+1)^2 \\ y_h &= (x_1^2 + x_2 - 11)^2 + (x_2^2 + x_1 - 7)^2, x_1, x_2 \in [-3, 3] \end{aligned} \tag{4.7}$$

Problem 2 (P2)

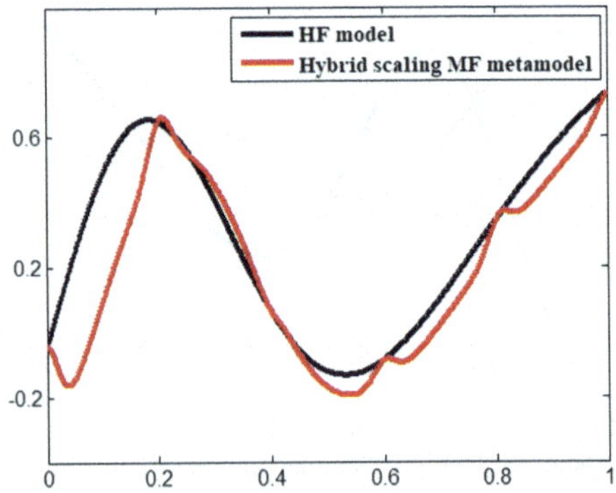

Fig. 4.4 MF surrogate model with hybrid scaling for the 1D function

Table 4.1 Features of the numerical test problems

Problem number	Problem scale	Nonlinearity order
P1	Low ($N_V = 2$)	Low
P2	Low ($N_V = 4$)	High
P3	High ($N_V = 8$)	Low
P4	High ($N_V = 10$)	High

$$\begin{aligned} y_l &= (x_1-1)^2 + 2\times(2x_2^2-0.75x_1)^2 + 3\times(3x_3^2-0.75x_2)^2 + 4\times(4x_4^2-0.75x_3)^2 \\ y_h &= (x_1-1)^2 + 2\times(2x_2^2-x_1)^2 + 3\times(3x_3^2-x_2)^2 + 4\times(4x_4^2-x_3)^2 \\ & x_1, x_2, x_3, x_4 \in [-10, 10] \end{aligned} \tag{4.8}$$

Problem 3 (P3)

$$\begin{aligned} f_{borehole} &= \frac{2\pi x_3(x_4-x_6)}{\ln(x_2/x_1)[1+2x_7x_4/(\ln(x_2/x_1)x_1^2x_8)+x_3/x_5]} \\ y_l &= 0.4f_{borehole}(x) + 0.07x_1^2x_8 + x_1x_7/x_3 + x_1x_6/x_2 + x_1^2x_4 \\ y_h &= f_{borehole}(x) \\ & x_1 \in [0.05, 0.15],\ x_2 \in [100, 50000], \\ & x_3\ [63070, 115600],\ x_4 \in [990, 1110], \\ & x_5 \in [63.1, 116],\ x_6 \in [700, 820], \\ & x_7 \in [1120, 1680],\ x_8 \in [9855, 12045] \end{aligned} \tag{4.9}$$

Problem 4 (P4)

$$\begin{aligned} y_l &= \sum_{i=1}^{10} x_i^3 + (\sum_{i=1}^{10} 2ix_i)^2 + (\sum_{i=1}^{10} 3ix_i)^4 \\ y_h &= \sum_{i=1}^{10} x_i^2 + (\sum_{i=1}^{10} 0.5ix_i)^2 + (\sum_{i=1}^{10} 0.5ix_i)^4,\ -5 \le x_i \le 10 \end{aligned} \tag{4.10}$$

Two surrogate model accuracy metrics are used in this section to compare the different approaches, namely, the relative maximum absolute error (RMAE) and the relative root mean square error (RRMSE), which are defined below:

$$
\begin{aligned}
\text{RMAE} &= \frac{1}{\text{STD}}\sqrt{\frac{1}{N}\sum_{i=1}^{N}(y_i - \hat{y}_i)^2} \\
\text{RRMSE} &= \frac{1}{\text{STD}}\max(|y_i - \hat{y}_i|) \\
\text{STD} &= \sqrt{\frac{1}{N-1}\sum_{i=1}^{N}(y_i - \bar{y})^2}
\end{aligned}
\tag{4.11}
$$

where N is the total number of validation points, $\bar{y}$ is the mean of the observed values at the validation points, and y_i and $\hat{y}_i$ are the real response and the predicted value, respectively, at the i-th validation point. The lower the value of the RMAE/ RRMSE is, the more accurate the MF surrogate model. The RRMSE is used to gauge the overall accuracy of the model, while the RMAE is used to gauge the local accuracy of the model. An additional 100 randomly selected sample points are used to calculate the RMAE and RRMSE in this section.

To construct a variable-fidelity surrogate model, sample points at two levels (HF and LF) should first be generated. Here, to ensure good uniformity of the samples and flexibility of the sample size, the Latin hypercube sampling (LHS) method (Huntington and Lyrintzis 1998) was chosen to generate the initial set of samples within the design domain. The samples were obtained using the MATLAB 2013b routine 'lhsdesign' with the 'maximum' criterion, which maximizes the minimum distance between sample points.

Two circumstances are considered for the comparison of the different approaches: large HF sample sets (N_{HF} = 12d) and small HF sample sets (N_{HF} = 5d). The number of LF samples is 20d in both cases, where d is the dimensionality of the test problem. The four numerical problems were tested 30 times for each of the different approaches to account for the influence of randomness.

In Table 4.2, the best error metrics are marked in bold. For problems with different features and different sizes, the three scaling-function-based approaches show different performances. No approach is universally better than the other approaches for all problems; therefore, it is important to choose the most suitable approach based on the characteristics of the problem of interest.

4.2 Space Mapping (SM) Approaches

4.2.1 The Concept of Output-Output Space Mapping (OOSM) Approaches

The goal of an output-output space mapping (OOSM) approach is to construct an MF surrogate model by taking the LF output values as prior knowledge of the studied system and directly mapping them to the output of the HF model.

Table 4.2 Test results obtained for the numerical examples using different approaches

Sample size	Examples	Error metric	Multiplicative scaling approach		Additive scaling approach		Hybrid scaling approach	
			Mean	STD	Mean	STD	Mean	STD
Small size	P1	RMAE	1.0515	0.3913	1.0753	0.2872	**1.0076**	0.3318
		RRMSE	0.3778	0.1641	0.3797	0.1250	**0.3555**	0.1379
	P2	RMAE	1.1576	0.9012	**0.7042**	0.3154	0.8790	0.4553
		RRMSE	0.2976	0.2675	**0.1734**	0.0509	0.2142	0.1246
	P3	RMAE	**0.0140**	0.0040	0.0205	0.0059	0.0171	0.0048
		RRMSE	**0.0032**	0.0006	0.0048	0.0009	0.0039	0.0007
	P4	RMAE	**3.7392**	8.2381	1293.34	291.48	646.67	145.64
		RRMSE	**0.6089**	0.9223	436.85	77.3522	218.43	38.65
Large size	P1	RMAE	0.4570	0.2058	**0.2767**	0.1543	0.3339	0.1729
		RRMSE	0.0967	0.0437	**0.0572**	0.0342	0.0705	0.0355
	P2	RMAE	2.2591	2.8564	**0.5497**	0.1776	1.2563	1.3664
		RRMSE	0.5524	0.8175	**0.1442**	0.0309	0.3159	0.3985
	P3	RMAE	**0.0101**	0.0046	0.0121	0.0052	0.0111	0.0049
		RRMSE	**0.0021**	0.0004	0.0024	0.0005	0.0022	0.0004
	P4	RMAE	**3.7392**	8.2381	1293.34	291.48	646.67	145.64
		RRMSE	**0.6089**	0.9223	436.85	77.3522	218.43	38.6578

A schematic diagram is presented in Fig. 4.5 to illustrate the similarities and differences between scaling-function-based MF approaches and the OOSM approach.

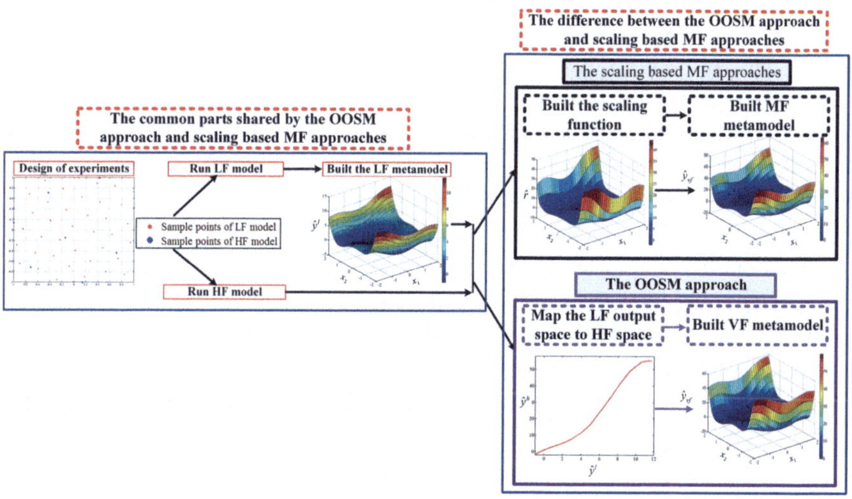

Fig. 4.5 Comparison between the OOSM approach and the scaling-function-based MF approaches

As seen from Fig. 4.5, both the scaling-function-based MF approaches and the OOSM approach use the LF model to capture the overall trend of the characteristics of the system. In scaling-function-based MF approaches, surrogate models for the scaling function are constructed to learn the differences between the HF responses and the predicted output values from the LF surrogate model. This is actually a process of mapping a multidimensional space to a 1D space, which is expected to yield a significantly higher prediction accuracy than can be obtained using the single HF surrogate model with a small amount of HF data for problems in which the relationship between the difference response features of the HF and LF models and the design variables is simple (Zheng et al. 2014). By contrast, in the OOSM approach, a surrogate model is constructed by mapping a 1D space (the output space of the LF model) to another 1D space (the output space of the HF model). Clearly, the OOSM approach will alleviate the computational burden of MF modelling when the dimensionality of the design space is greater than one.

4.2.2 The Radial Basis Function (RBF)-Based OOSM Approach

In this section, the formulation and derivation processes for the radial basis function (RBF)-based OOSM approach are presented along with its framework. Figure 4.6 illustrates the framework of the RBF-based OOSM approach. Throughout this section, we will review the key points involved in each step.

Step 1: Generate two sample sets, $\boldsymbol{X}^l$ and $\boldsymbol{X}^h$

Since the LF model is able to reflect the most prominent features of the system at a considerably reduced computational cost, a sample set $\boldsymbol{X}^l = \{\boldsymbol{x}_1^l, \boldsymbol{x}_2^l, \ldots, \boldsymbol{x}_N^l\}$ with a relatively large number of sample points is generated to ensure the accuracy of this model. By contrast, a sample set $\boldsymbol{X}^h = \{\boldsymbol{x}_1^h, \boldsymbol{x}_2^h, \ldots, \boldsymbol{x}_M^h\}$ with a significantly smaller number of sample points is generated to obtain the response values of the HF model. In this step, the optimal Latin hypercube sampling (OLHS) method (Jin et al. 2005), which can ensure that the design space is uniformly covered, is applied to generate the sample sets for the LF and HF models.

Step 2: Run the LF model on sample set $\boldsymbol{X}^l$ to obtain the LF response values

Based on the sample set $\boldsymbol{X}^l$ generated in Step 1, the actual LF response vector $\boldsymbol{f}^l = \{f_1^l, f_2^l, \ldots, f_N^l\}$ is obtained by running the LF model.

Step 3: Build an RBF surrogate model for the LF model

Based on the LF sampling data, an RBF surrogate model $\hat{f}^l(\boldsymbol{x})$ is built for the LF model. Then, the value predicted by the LF model at any point $\boldsymbol{x}$ can be expressed as

$$\hat{f}^l(\boldsymbol{x}) = \sum_{p=1}^{N} w_p^l \phi(\left\|\boldsymbol{x} - \boldsymbol{x}_p^l\right\|) \tag{4.12}$$

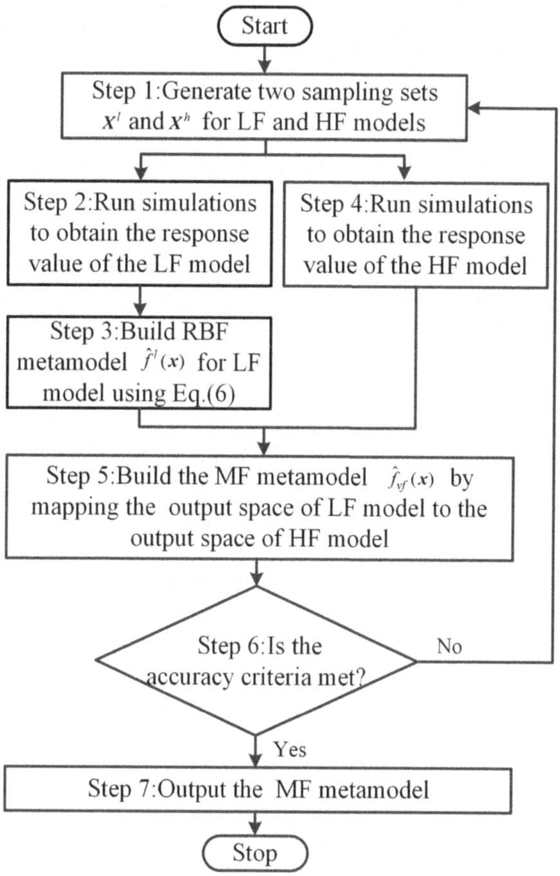

Fig. 4.6 Framework of the RBF-based OOSM approach

where N is the number of sample points for the LF model, $\boldsymbol{x}_p^l$ is the p-th sample point in $\boldsymbol{X}^l$, $\boldsymbol{x}$ is the design variable in the design space and $\left\|\boldsymbol{x}-\boldsymbol{x}_p^l\right\|$ represents the Euclidean distance between the design variable and the p-th LF sample point and is mathematically expressed as

$$\left\|\boldsymbol{x}-\boldsymbol{x}_p^l\right\| = \sqrt{(\boldsymbol{x}-\boldsymbol{x}_p^l)^T(\boldsymbol{x}-\boldsymbol{x}_p^l)} \tag{4.13}$$

$\phi(\cdot)$ represents the RBFs. Commonly used RBFs are listed as follows (Zhou and Jiang 2016):

(1) Bi-harmonic, $\phi(r) = r$; (2) Thin-plate spline, $\phi(r) = r^2\log(r)$;
(3) Multi-quadric, $\phi(r) = \sqrt{r^2+c^2}$; (4) Cubic, $\phi(r) = (r+c)^3$;
(5) Gaussian, $\phi(r) = e^{-(cr^2)}$; (6) Inverse multi-quadric, $\phi(r) = \frac{1}{\sqrt{r^2+c^2}}$,

where c is a constant value and $0 < c \leq 1$.

The unknown interpolation vector w^l is obtained by minimizing the sum of the squares of the deviations, which can be expressed as

$$J_{lf} = \sum_{s=1}^{N} \left[f(\mathbf{x}_s^l) - \sum_{p=1}^{N} w_p^l \phi(\left\| \mathbf{x}_s^l - \mathbf{x}_p^l) \right\|) \right]^2 \tag{4.14}$$

By solving the above optimization problem, the coefficients w^l can be obtained as follows:

$$w^l = \left(\Phi_l^T \Phi_l + \Lambda_l \right)^{-1} \Phi_l^T \mathbf{f}^l \tag{4.15}$$

where the elements of Λ_l are all zero except for the regularization parameters along the diagonal and Φ_l is the design matrix. Because of its advantages of relatively few parameters to be set and excellent overall performance, the Gaussian form is adopted for the RBFs. Then, the design matrix Φ_l can be obtained as follows:

$$\Phi_l = \begin{pmatrix} e^{-(c \cdot \left\| \mathbf{x}_1^l - \mathbf{x}_1^l \right\|)^2} & e^{-(c \cdot \left\| \mathbf{x}_1^l - \mathbf{x}_2^l \right\|)^2} & \cdots & e^{-(c \cdot \left\| \mathbf{x}_1^l - \mathbf{x}_N^l \right\|)^2} \\ e^{-(c \cdot \left\| \mathbf{x}_2^l - \mathbf{x}_1^l \right\|)^2} & e^{-(c \cdot \left\| \mathbf{x}_2^l - \mathbf{x}_2^l \right\|)^2} & \cdots & e^{-(c \cdot \left\| \mathbf{x}_2^l - \mathbf{x}_N^l \right\|)^2} \\ \vdots & \vdots & \ddots & \vdots \\ e^{-(c \cdot \left\| \mathbf{x}_N^l - \mathbf{x}_1^l \right\|)^2} & e^{-(c \cdot \left\| \mathbf{x}_N^l - \mathbf{x}_2^l \right\|)^2} & \cdots & e^{-(c \cdot \left\| \mathbf{x}_N^l - \mathbf{x}_N^l \right\|)^2} \end{pmatrix} \tag{4.16}$$

Step 4: Run the HF model on the sample set $\mathbf{X}^h$ to obtain the HF response values

Based on the sample set $\mathbf{X}^h$ generated in Step 1, the actual HF response vector $\mathbf{f}^h = \{f_1^h, f_2^h, \ldots, f_M^h\}$ is obtained by running the HF model.

Step 5: Build the MF surrogate model by taking the LF predicted values as prior knowledge and directly mapping them to the output of the HF model.

By taking the constructed LF surrogate model $\hat{f}^l(\mathbf{x})$ as prior knowledge and mapping its output space to that of the HF model, the formulation for the proposed OOSM model can be specified as a linear combination of RBFs with weight coefficients, as shown in the following equation:

$$\begin{aligned} \hat{f}_{MF}(\mathbf{x}) &= \sum_{q=1}^{M} w_q^{OSM} \phi(\hat{f}^l(\mathbf{x}), \hat{f}^l(\mathbf{x}_q^h)) \\ &= \sum_{q=1}^{M} w_q^{OSM} \phi(\left\| \hat{f}^l(\mathbf{x}) - \hat{f}^l(\mathbf{x}_q^h) \right\|) \end{aligned} \tag{4.17}$$

Here, M is the number of sample points for the HF model. $\mathbf{x}_q^h$ is the q-th sample point in $\mathbf{X}^h$, $\hat{f}^l(\mathbf{x})$ is the predicted value at the design variable $\mathbf{x}$ as obtained from the

LF surrogate model, which can be calculated using Eq. (4.12). Similarly, $\hat{f}^l(\boldsymbol{x}_q^h)$ is the value at $\boldsymbol{x}_q^h$ predicted by the LF surrogate model, which can be calculated as

$$\hat{f}^l(\boldsymbol{x}_q^h) = \sum_{p=1}^{N} w_p^l \phi(\left\|\boldsymbol{x}_q^h - \boldsymbol{x}_p^l\right\|) \tag{4.18}$$

where $\left\|\boldsymbol{x}_q^h - \boldsymbol{x}_p^l\right\|$ represents the Euclidean distance between the LF and HF sample points, which can be expressed as

$$\left\|\boldsymbol{x}_q^h - \boldsymbol{x}_p^l\right\| = \sqrt{(\boldsymbol{x}_q^h - \boldsymbol{x}_p^l)^T(\boldsymbol{x}_q^h - \boldsymbol{x}_p^l)} \tag{4.19}$$

The unknown interpolation coefficient w^{OSM} is obtained by minimizing the sum of the squares of the deviations, which can be expressed as

$$J_{MF} = \sum_{k=1}^{M}\left[f^h(\boldsymbol{x}_k^h) - \sum_{q=1}^{M} w_q^{OSM}\phi(\left\|\hat{f}^l(\boldsymbol{x}_k^h) - \hat{f}^l(\boldsymbol{x}_q^h)\right\|)\right]^2 \tag{4.20}$$

When a weight penalty term is added to the sum of the squares of the deviations, the cost function is minimized as follows:

$$C_{MF} = \sum_{k=1}^{M}\left[f^h(\boldsymbol{x}_k^h) - \sum_{q=1}^{M} w_q^{OSM}\phi(\left\|\hat{f}^l(\boldsymbol{x}_k^h) - \hat{f}^l(\boldsymbol{x}_q^h)\right\|)\right]^2 + \sum_{q=1}^{M}\lambda_q(w_q^{OSM})^2 \tag{4.21}$$

where λ is a nonnegative regularization vector that is used to control the additional weight penalty term.

To solve the above optimization problem to obtain the coefficients w^{OSM}, the partial derivatives of the cost function with respect to each w_q^{OSM} are calculated. The partial derivative with respect to the q-th coefficient can be expressed as

$$\begin{aligned}\frac{\partial C_{vf}}{\partial w_q^{OSM}} &= 2\sum_{k=1}^{M}(w_q^{OSM}\phi(\left\|\hat{f}^l(\boldsymbol{x}_k^h) - \hat{f}^l(\boldsymbol{x}_q^h)\right\|) - f^h(\boldsymbol{x}_k^h))\frac{\partial(w_q^{OSM}\phi(\left\|\hat{f}^l(\boldsymbol{x}_k^h) - \hat{f}^l(\boldsymbol{x}_q^h)\right\|))}{\partial w_q^{OSM}}) + 2\lambda_q w_q^{OSM} \\ &= 2\sum_{k=1}^{M}(w_q^{OSM}\phi(\left\|\hat{f}^l(\boldsymbol{x}_k^h) - \hat{f}^l(\boldsymbol{x}_q^h)\right\|) - f^h(\boldsymbol{x}_k^h))\phi(\left\|\hat{f}^l(\boldsymbol{x}_k^h) - \hat{f}^l(\boldsymbol{x}_q^h)\right\|)) + 2\lambda_q w_q^{OSM}\end{aligned} \tag{4.22}$$

Setting the above expression equal to zero leads to the following equation:

$$\sum_{k=1}^{M}(w_q^{OSM}\phi(\left\|\hat{f}^l(\boldsymbol{x}_k^h)-\hat{f}^l(\boldsymbol{x}_q^h)\right\|)\phi(\left\|\hat{f}^l(\boldsymbol{x}_k^h)-\hat{f}^l(\boldsymbol{x}_q^h)\right\|))+\lambda_q w_q^{OSM}$$
$$=\sum_{k=1}^{M}(f^h(\boldsymbol{x}_k^h)\phi(\left\|\hat{f}^l(\boldsymbol{x}_k^h)-\hat{f}^l(\boldsymbol{x}_q^h)\right\|)) \tag{4.23}$$

There are M such equations, for $1\leq q\leq M$, each representing one constraint on the solution. When matrices and vectors are used, the problem for obtaining w^{OSM} can be rewritten as

$$\Phi_{MF}^T\Phi_{MF}w^{OSM}+\Lambda w^{OSM}=\Phi_{MF}^T\boldsymbol{f}^h \tag{4.24}$$

where Φ_{MF} is the design matrix. When the Gaussian form is adopted for the RBFs, the design matrix Φ_{MF} can be obtained as follows:

$$\Phi_{MF}=\begin{pmatrix} e^{-(c\cdot\left\|\hat{f}^l(x_1^h)-\hat{f}^l(x_1^h)\right\|)^2} & e^{-(c\cdot\left\|\hat{f}^l(x_1^h)-\hat{f}^l(x_2^h)\right\|)^2} & \cdots & e^{-(c\cdot\left\|\hat{f}^l(x_1^h)-\hat{f}^l(x_M^h)\right\|)^2} \\ e^{-(c\cdot\left\|\hat{f}^l(x_2^h)-\hat{f}^l(x_1^h)\right\|)^2} & e^{-(c\cdot\left\|\hat{f}^l(x_2^h)-\hat{f}^l(x_2^h)\right\|)^2} & \cdots & e^{-(c\cdot\left\|\hat{f}^l(x_2^h)-\hat{f}^l(x_M^h)\right\|)^2} \\ \vdots & \vdots & \ddots & \vdots \\ e^{-(c\cdot\left\|\hat{f}^l(x_M^h)-\hat{f}^l(x_1^h)\right\|)^2} & e^{-(c\cdot\left\|\hat{f}^l(x_M^h)-\hat{f}^l(x_2^h)\right\|)^2} & \cdots & e^{-(c\cdot\left\|\hat{f}^l(x_M^h)-\hat{f}^l(x_M^h)\right\|)^2} \end{pmatrix} \tag{4.25}$$

The elements of Λ are all zero except for the regularization parameters along the diagonal:

$$\Lambda=\begin{pmatrix} \lambda_1 & 0 & \cdots & 0 \\ 0 & \lambda_2 & \cdots & 0 \\ \vdots & \vdots & \ddots & \vdots \\ 0 & 0 & \cdots & \lambda_m \end{pmatrix} \tag{4.26}$$

By solving Eq. (4.18), the coefficients w^{OSM} can be obtained as follows:

$$w^{OSM}=\left(\Phi_{MF}^T\Phi_{MF}+\Lambda\right)^{-1}\Phi_{MF}^T(\boldsymbol{f}^h)^T \tag{4.27}$$

Finally, by substituting Eqs. (4.12), (4.18) and (4.27) into Eq. (4.17), the predicted values at any design points as obtained from the proposed OOSM model can be calculated as follows:

$$\hat{f}_{MF}(\boldsymbol{x})=\sum_{q=1}^{M}w_q^{OSM}\phi(\left\|\sum_{p=1}^{N}w_p^l\phi(\left\|\boldsymbol{x}-\boldsymbol{x}_p^l\right\|)-\sum_{p=1}^{N}w_p^l\phi(\left\|\boldsymbol{x}_q^h-\boldsymbol{x}_p^l\right\|)\right\|) \tag{4.28}$$

Step 6: Check the prediction accuracy of the obtained MF surrogate model

The prediction performance of the MF surrogate model created through the above steps needs to be validated before it can be used in support of simulation-based design. If the preselected prediction accuracy criterion is not achieved, the process will return to Step 1; otherwise, it will proceed to Step 7.

Step 7: Output the final MF surrogate model

Once the preselected prediction accuracy of the obtained MF surrogate model has been achieved, the algorithm will output the final MF surrogate model.

A numerical example adapted from Zhou et al. (2015) will be used here to present a detailed comparison among different multi-fidelity modelling (MFM) approaches. The mathematical description of the numerical example is given below:

Modified six-hump camelback (MSC) function:

$$\begin{aligned} f(x_1,x_2) &= 4x_1^2 - 2.1x_1^4 + x_1^6/3 + x_1x_2 - 4x_2^2 + 4x_2^4; \\ f^h &= f(x_1,x_2); \\ f^l &= f(0.7x_1, 0.7x_1); \\ x_1 &\in [-2,\ 2], x_2 \in [-2,\ 2] \end{aligned} \tag{4.29}$$

where f^h denotes the HF model that is to be approximated and f^l denotes the LF model. Note that the analytical functions are used only to obtain the response values at a given sample point. Although these analytical functions are explicit, the overall relationships between the input variables and the corresponding LF and HF responses, which are equivalent to the relationship between the HF and LF models, are assumed to be unknown.

The OLHS method proposed by Jin et al. (2005) was used to generate sample sets. Three different sample sizes for the HF model, labelled sample size 1 ($M = 2d + 1$), sample size 2 ($M = 3d + 2$) and sample size 3 ($M = 4d + 3$) in Fig. 4.7, are considered, where d is the dimensionality of the design space.

To calculate the values of the two prediction accuracy metrics for the different MFM approaches with different sample sizes, an additional 1024 test points were

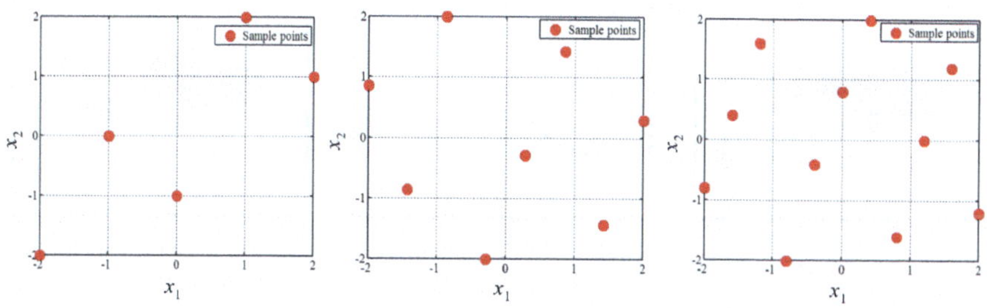

(a) Sample size 1 (N=2d+1) (b) Sample size 2 (N=3d+2) (c) Sample size 3 (N=4d+3)

Fig. 4.7 Three sample sets of different sizes generated via OLHS randomly selected. For each MFM approach, 10 different runs were performed for the numerical example of the MSC function, and the average values of the two prediction accuracy metrics across these runs are presented as the final results to avoid unrepresentative numerical results. Figure 4.8a plots the actual HF and LF models for the MSC function, and a contour map of their differences is presented in Fig. 4.8b.

Figure 4.9 illustrates the MF surrogate model constructed by mapping the output space of the LF model to the output space of the HF model, a 1D-to-1D mapping process. Figure 4.9c illustrates the final MF surrogate model obtained using the RBF-based OOSM approach. By comparing Fig. 4.9c with Fig. 4.8a, it can be concluded that the MF surrogate model constructed with the proposed OOSM approach is capable of describing the behaviour of the actual HF model with high accuracy over the whole design domain.

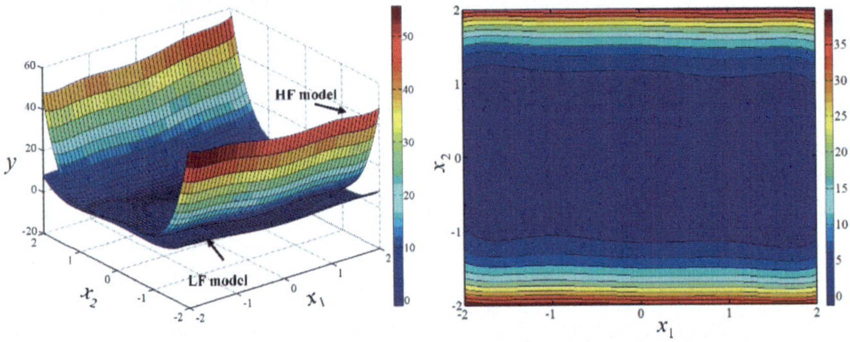

(a) Actual HF model and LF model for MSC

(b) Coutours of difference between HF and LF models for MSC

Fig. 4.8 Plots of the actual HF and LF models for the MSC function

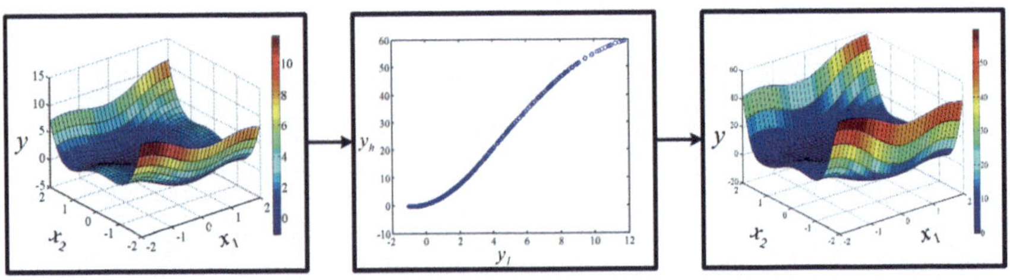

(a) The LF metamodel

(b) Low-fidelity output space mapping process

(c) MF metamodel obtained by OOSM method for MSC

Fig. 4.9 Numerical illustration of the RBF-based OOSM approach

4.2.3 The Gaussian Process-Based OOSM Approach

The aim of the SM-based MFM approach proposed in this section is to address the limitations of scaling-function-based MFM approaches by treating the LF output information as prior knowledge of the studied system and directly mapping it to the output space of the HF model using a GP model. Based on the available LF and HF information, as shown in Eqs. (4.30) and (4.31), the response at an unobserved point $\boldsymbol{x}$ as predicted using the proposed GM-based OOSM approach can be expressed as a realization of a regression model F and a stochastic process z,

$$f_{MF}(\boldsymbol{x}) = F(\boldsymbol{\beta}, \hat{f}^l(\boldsymbol{x})) + z(\hat{f}^l(\boldsymbol{x})) \tag{4.30}$$

where $\hat{f}^l(\boldsymbol{x})$ denotes the response at $\boldsymbol{x}$ predicted by the LF surrogate model, which is constructed using procedure used in the scaling-function-based MFM approaches. Readers are referred to (Han and Görtz 2012; Zhou et al. 2016a) for the details of building such an LF surrogate model.

$F(\boldsymbol{\beta}, \hat{f}^l(\boldsymbol{x}))$ is a linear combination of p chosen functions:

$$\begin{aligned} F(\boldsymbol{\beta}, \hat{f}^l(\boldsymbol{x})) &= \beta_1 h_1(\hat{f}^l(\boldsymbol{x})) + \beta_2 h_2(\hat{f}^l(\boldsymbol{x})) + \cdots \beta_p h_p(\hat{f}^l(\boldsymbol{x}) \\ &= \left[h_1(\hat{f}^l(\boldsymbol{x})), h_2(\hat{f}^l(\boldsymbol{x})), \ldots, h_p(\hat{f}^l(\boldsymbol{x})\right] \boldsymbol{\beta} \\ &= \boldsymbol{h}(\hat{f}^l(\boldsymbol{x}))^T \boldsymbol{\beta} \end{aligned} \tag{4.31}$$

where $\boldsymbol{\beta}$ denotes a column vector of regression coefficients and $\boldsymbol{h}(\hat{f}^l(\boldsymbol{x}))$ denotes a row vector of regression functions. The proposed approach for selecting the optimal regression function is described in Sect. 3.1.1.

The stochastic process z is assumed to have a mean of zero and a covariance between $z(\hat{f}^l(\boldsymbol{x}))$ and $z(\hat{f}^l(\boldsymbol{x}'))$,

$$E\left[z(\hat{f}^l(\boldsymbol{x})), z(\hat{f}^l(\boldsymbol{x}'))\right] = \sigma^2 R(\boldsymbol{\theta}, \hat{f}^l(\boldsymbol{x}), \hat{f}^l(\boldsymbol{x}')) \tag{4.32}$$

where σ is the process standard deviation determining the overall magnitude of the variance and $R(\boldsymbol{\theta}, \hat{f}^l(\boldsymbol{x}), \hat{f}^l(\boldsymbol{x}'))$ is the correlation model. In this work, the most popular form of the correlation function, a Gaussian exponential function (Kleijnen 2017), is adopted:

$$R(\theta, \hat{f}^l(\boldsymbol{x}), \hat{f}^l(\boldsymbol{x}')) = \prod_{d=1}^{d} \exp(-\theta_d |\hat{f}^l(x_d) - \hat{f}^l(x'_d)|^2) \tag{4.33}$$

where d is the dimensionality of the design variable and the elements of $\boldsymbol{\theta}$ are the roughness parameters. The solution method for $\boldsymbol{\theta}$ is described in Sect. 3.1.2.

It is assumed that the response of the MF surrogate model can be approximated as a linear combination of the HF information:

$$\hat{f}_{MF}(\hat{f}^l(\boldsymbol{x})) = \boldsymbol{c}^T \boldsymbol{f}^h \tag{4.34}$$

where $\boldsymbol{c} = [c^1, \ldots, c^M]$ is a vector of weight coefficients associated with the available HF data.

For a set $S = [\hat{f}^l(\boldsymbol{x}_1^h), \hat{f}^l(\boldsymbol{x}_2^h), \ldots, \hat{f}^l(\boldsymbol{x}_M^h)]$, the expanded $M \times p$ design matrix F is defined as

$$F = \left[\boldsymbol{h}(\hat{f}^l(\boldsymbol{x}_1^h)), \boldsymbol{h}(\hat{f}^l(\boldsymbol{x}_2^h)), \ldots, \boldsymbol{h}(\hat{f}^l(\boldsymbol{x}_M^h))\right]^T \tag{4.35}$$

Furthermore, the matrix R of the stochastic process correlations between the z values at different design points is defined as follows:

$$R_{ij} = R(\theta, \hat{f}^l(\boldsymbol{x}_i^h), \hat{f}^l(\boldsymbol{x}_j^h)) \qquad i, j = 1, 2, \ldots, M \tag{4.36}$$

For an unobserved point $\boldsymbol{x}$, the vector of correlations between it and the z values at the design points is defined as

$$r(\hat{f}^l(\boldsymbol{x})) = \left[R(\theta, \hat{f}^l(\boldsymbol{x}_1^h), \hat{f}^l(\boldsymbol{x})), R(\theta, \hat{f}^l(\boldsymbol{x}_2^h), \hat{f}^l(\boldsymbol{x})), \ldots, R(\theta, \hat{f}^l(\boldsymbol{x}_M^h), \hat{f}^l(\boldsymbol{x}))\right]^T \tag{4.37}$$

Then, the errors between the linear predictor $\hat{f}_{vf}(\hat{f}^l(\boldsymbol{x}))$ and $f_{vf}(\hat{f}^l(\boldsymbol{x}))$ at an unobserved point $\boldsymbol{x}$ can be calculated as follows:

$$\begin{aligned} \hat{f}_{MF}(\hat{f}^l(\boldsymbol{x})) - f_{MF}(\hat{f}^l(\boldsymbol{x})) &= \boldsymbol{c}^T \boldsymbol{f}^h - (\boldsymbol{h}(\hat{f}^l(\boldsymbol{x}))^T \boldsymbol{\beta} + z) \\ &= \boldsymbol{c}^T (F\boldsymbol{\beta} + \boldsymbol{Z}) - ((\boldsymbol{h}(\hat{f}^l(\boldsymbol{x}))^T \boldsymbol{\beta} + z) \\ &= \boldsymbol{c}^T \boldsymbol{Z} - z + (F^T \boldsymbol{c} - \boldsymbol{h}(\hat{f}^l(\boldsymbol{x})))^T \boldsymbol{\beta} \end{aligned} \tag{4.38}$$

where $\boldsymbol{Z} = [z_1, z_2, \ldots, z_M]$ denotes the errors at the design points. To keep the predictor unbiased, it is required that

$$F^T \boldsymbol{c} - \boldsymbol{h}(\hat{f}^l(\boldsymbol{x})) = 0 \tag{4.39}$$

Under this condition, the mean square error (MSE) of the predictor in Eq. (4.38) is

$$\begin{aligned}\varphi(x) &= E[(\hat{f}_{MF}(\hat{f}^l(\boldsymbol{x})) - f_{MF}(\hat{f}^l(\boldsymbol{x})))^2] \\ &= E[(\boldsymbol{c}^T Z - z)^2] \\ &= E[z^2 + \boldsymbol{c}^T Z Z^T \boldsymbol{c} - 2\boldsymbol{c}^T Z z] \\ &= \sigma^2(1 + \boldsymbol{c}^{\mathrm{T}} R c - 2\boldsymbol{c}^T r)\end{aligned} \tag{4.40}$$

To minimize the MSE, a Lagrange multiplier λ can be introduced, and the Lagrangian function with respect to $\boldsymbol{c}$ is

$$L(\boldsymbol{c}, \lambda) = \sigma^2(1 + \boldsymbol{c}^T R c - 2\boldsymbol{c}^T r) - \lambda^T(F^T \boldsymbol{c} - \boldsymbol{h}(\hat{f}^l(\boldsymbol{x}))) \tag{4.41}$$

The gradient of Eq. (4.41) with respect to c can be calculated as follows:

$$L'_c(\boldsymbol{c}, \lambda) = 2\sigma^2(R\boldsymbol{c} - r) - F\lambda \tag{4.42}$$

From the first-order necessary conditions for optimality, the following system of equations can be obtained:

$$\begin{bmatrix} R & F \\ F^T & 0 \end{bmatrix} \begin{bmatrix} c \\ \tilde{\lambda} \end{bmatrix} = \begin{bmatrix} r \\ \boldsymbol{h}(\hat{f}^l(\boldsymbol{x})) \end{bmatrix} \tag{4.43}$$

where $\tilde{\lambda}$ is defined as $\tilde{\lambda} = -\frac{\lambda}{2\sigma^2}$.

Then, solving Eq. (4.43) yields

$$\begin{aligned}\tilde{\lambda} &= (F^T R^{-1} F)^{-1}(F^T R^{-1} r - \boldsymbol{h}(\hat{f}^l(\boldsymbol{x}))) \\ c &= R^{-1}(r - F\tilde{\lambda})\end{aligned} \tag{4.44}$$

Because the matrix R is symmetric, upon substituting Eq. (4.44) into Eq. (4.34), the predictor obtained via the proposed GP-based OOSM approach for any unobserved point $\boldsymbol{x}$ is given by

$$\begin{aligned}\hat{f}_{vf}(\boldsymbol{x}) &= (r - F\tilde{\lambda})^T R^{-1} \boldsymbol{f}^h \\ &= r^T R^{-1} \boldsymbol{f}^h - (F^T R^{-1} r - \boldsymbol{h}(\hat{f}^l(\boldsymbol{x})))^T (F^T R^{-1} F)^{-1} F^T R^{-1} \boldsymbol{f}^h\end{aligned} \tag{4.45}$$

To obtain the hyper-parameters $\boldsymbol{\theta}$, $\boldsymbol{\beta}$ and σ^2, the maximum likelihood estimates are calculated. The log-likelihood function is usually adopted for numerical purposes (Kleijnen 2009):

$$\begin{aligned} In(p(\boldsymbol{\beta}, \sigma^2, \boldsymbol{\theta})) = -\frac{M}{2} In(2\pi) - \frac{M}{2} In(\sigma^2) - \frac{1}{2} In(|R|^{1/2}) - \frac{1}{2\sigma^2}(\boldsymbol{f^h} - F\boldsymbol{\beta})^T R^{-1} \\ (\boldsymbol{f^h} - F\boldsymbol{\beta})^T \end{aligned} \tag{4.46}$$

To determine the maximum likelihood estimates of $\boldsymbol{\beta}$ and σ^2, the derivatives of Eq. (4.46) with respect to $\boldsymbol{\beta}$ and σ^2 are set to zero. Thus, the maximum likelihood estimates of $\boldsymbol{\beta}$ and σ^2 are found to be

$$\hat{\boldsymbol{\beta}} = (F^T R^{-1} F)^{-1} F^T R^{-1} \boldsymbol{f}^h \tag{4.47}$$

$$\hat{\sigma}^2 = \frac{1}{M} (\boldsymbol{f}^h - F\hat{\boldsymbol{\beta}})^T R^{-1} (\boldsymbol{f}^h - F\hat{\boldsymbol{\beta}}) \tag{4.48}$$

The hyper-parameters $\boldsymbol{\theta} = (\theta_1, \ldots, \theta_d)$ in Eq. (4.32) influence the attenuation rate of the correlation function. Before maximizing Eq. (4.46), by substituting Eq. (4.47) and Eq. (4.48) into it, the following expression is obtained:

$$\max In(p(\boldsymbol{\beta}, \sigma^2, \boldsymbol{\theta})) = -\frac{M}{2} In(2\pi) - \frac{M}{2} In(\hat{\sigma}^2) - \frac{1}{2} In(|R|^{1/2}) - \frac{M}{2} \tag{4.49}$$

To solve the maximum likelihood estimation problem expressed in Eq. (4.49), the free MATLAB toolbox DACE can be adopted, which relies on a powerful stochastic algorithm based on the Hooke and Jeeves method (Lophaven et al. 2002a). The maximum likelihood problem is difficult to solve because of three main difficulties (Martin and Simpson 2005; Gano et al. 2006; Kleijnen 2008a): (a) the multimodality of the log-likelihood function, (b) the long ridges in the log-likelihood function and (c) the ill-conditioned correlation matrix. The first and second issues can be addressed by adopting a global stochastic optimization algorithm, for example, Martin and Simpson (2004) suggested the use of simulated annealing (SA). However, such an algorithm is much more computationally expensive than a gradient-based method. Therefore, the question of how to establish a trade-off between the global and local solutions and the function call expense is still worth studying. The last issue can be addressed by adding a small nugget effect (e.g. 10^{-6}) to the diagonal elements of the correlation matrix.

A numerical example adapted from Aute et al. (2013) will be used to illustrate how the GP-based OOSM approach works. In this illustrative example,

$$\begin{aligned} f^l(\boldsymbol{x}) &= \sin(x_1)\cos(x_2); \\ f^h(\boldsymbol{x}) &= 1.5\sin^2(0.5x_1)\cos^2(0.5x_1)\cos(x_2) + 3\sin(x_1)\cos(x_2) - 0.5; \\ x_1 &\in [-2,\ 2], x_2 \in [-2,\ 2] \end{aligned} \tag{4.50}$$

where $f^{\mathrm{h}}(\boldsymbol{x})$ denotes the HF model to be approximated, while $f^l(\boldsymbol{x})$ denotes the LF model. These two functions are graphically shown in Fig. 4.10. Note that although the analytical functions here are explicitly known, the general relationship between the LF and HF models is assumed to be unknown.

In this example, the number of sample points for the LF model is fixed to $N = 30d$, where d denotes the dimensionality of the design space. Three different sample sizes for the HF model, $M = 3d$, $M = 5d$ and $M = 6d$, are considered to

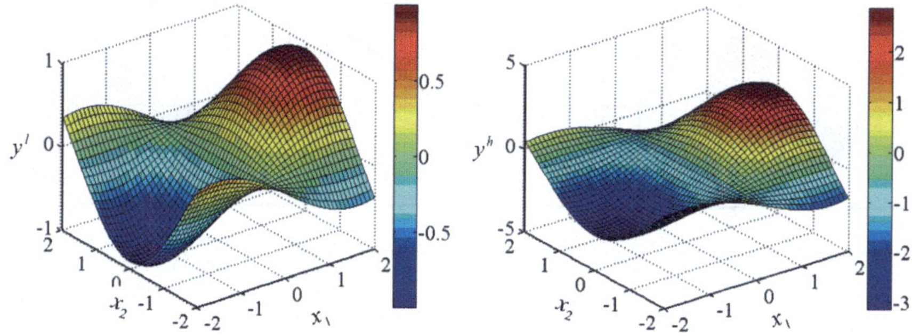

(a) Low-fidelity model for the numerical function

(b) High-fidelity model for the numerical function

Fig. 4.10 Plots of the LF and HF models

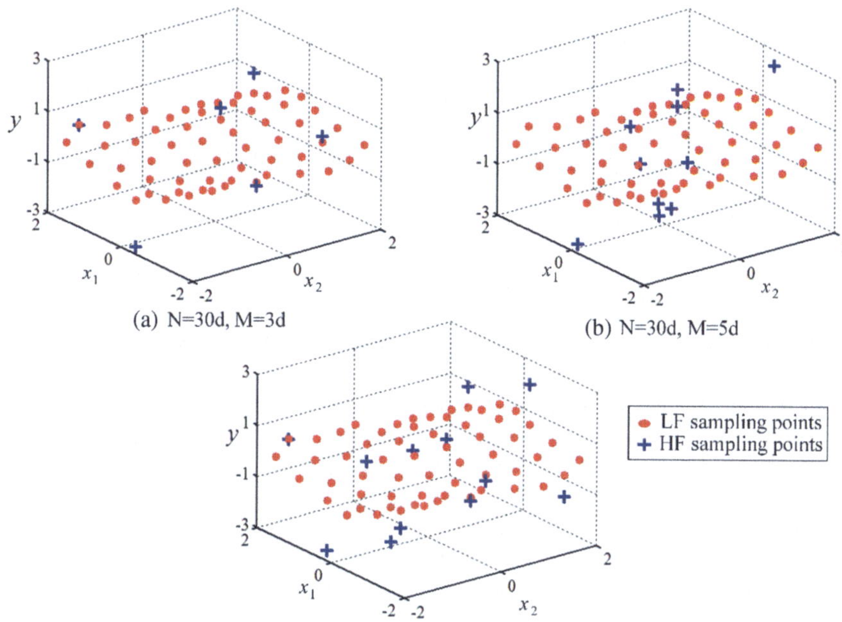

(a) N=30d, M=3d

(b) N=30d, M=5d

(c) N=30d, M=6d

Fig. 4.11 Sample sets with different HF sample sizes

study the prediction performance of different MFM approaches. In this work, OLHS (Park 1994) was used to ensure that the sampled points were spread throughout the design space. The generated sample points are plotted in Fig. 4.11.

Figure 4.12 illustrates the key steps of the GP-based OOSM approach for obtaining the MF surrogate model. As observed in Fig. 4.12, the LF information is

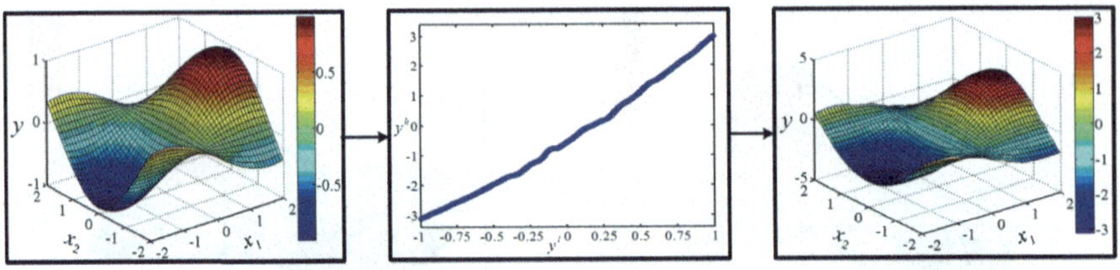

(a) The LF metamodel (b) Space mapping process (c) VF metamodel obtained by SM-VFM

Fig. 4.12 Numerical illustration of the GP-based OOSM approach

treated as an input and directly mapped to the output space of the HF model. This is actually a 1D-to-1D mapping process. The final obtained MF surrogate model is plotted in Fig. 4.12c. When the MF surrogate model in Fig. 4.12c is compared with the actual HF function plotted in Fig. 4.11b, it can be seen that the surrogate model constructed using the GP-based OOSM approach is able to accurately reflect the characteristics of the actual HF model over the whole design domain.

4.2.4 The Support Vector Regression (SVR)-Based OOSM Approach

In this section, a prior-knowledge least squares support vector regression (PKI-LSSVR) approach is developed to avoid the limitations of the commonly used difference mapping approach. The prior knowledge used here is the LF information, which can be derived from explicit empirical formulas or coarse black-box simulation models. Instead of mapping the differences between the LF and HF models, the PKI-LSSVR approach attempts to map the prior knowledge (the LF outputs) to the real HF outputs. The flowchart of PKI-LSSVR is illustrated in Fig. 4.13.

Consider a set of HF data $\{(x_{hi}, y_{hi})\}_{i=1,2,\ldots,m}$, with inputs $X_h = \{x_{h1}, x_{h2}, \ldots, x_{hm}\}$ and outputs $Y_h = \{y_{h1}, y_{h2}, \ldots, y_{hm}\}$, together with a much larger set of LF sample data $\{(x_{li}, y_{li})\}_{i=1,2,\ldots,n}$, where the inputs are $X_l = \{x_{l1}, x_{l2}, \ldots, x_{ln}\}$ and the corresponding outputs are $Y_l = \{y_{l1}, y_{l2}, \ldots, y_{ln}\}$. The purpose of SVR is to find a function that has the minimum prediction error for the training samples and the smallest deviation from the actual targets. Here, the LF outputs are treated as the inputs to LSSVR, and the relationship between the LF and HF outputs is then constructed via PKI-LSSVR, as follows:

$$\hat{F} = \boldsymbol{w}^T \varphi\left([\boldsymbol{X}_h \boldsymbol{Y}_l^h]\right) + b \tag{4.51}$$

where Y_l^h denotes the LF outputs at the HF sample points, which can be obtained by running the LF model $f_l(x)$ at X_h, that is, $Y_l^h = f_l(X_h)$ and $\varphi(\cdot)$ denotes a set of

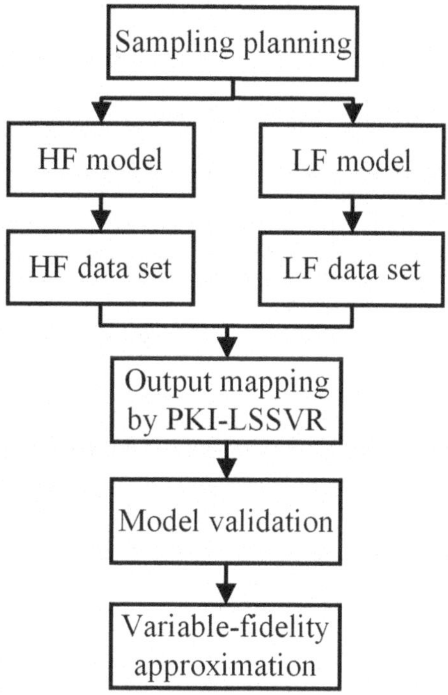

Fig. 4.13 Flowchart of the proposed PKI-LSSVR method

nonlinear transformations. Thus, the variable-fidelity surrogate model $\hat{F}$ is constructed by mapping the LF outputs to the HF outputs.

To construct the PKI-LSSVR model, the following optimization problem should be solved:

$$\min \frac{1}{2}\boldsymbol{w}^T\boldsymbol{w} + \frac{1}{2}C\sum_{i=1}^{m} e_i^2 \tag{4.52}$$

$$\text{s.t.} e_i = y_{hi} - \boldsymbol{w}^T\varphi\left(\left[\boldsymbol{X}_h\boldsymbol{Y}_l^h\right]\right) - b, i = 1, 2, \ldots, m \tag{4.53}$$

where m is the number of HF sample points.

The Lagrangian form of the above optimization problem can be expressed as follows:

$$\mathrm{L} = \frac{1}{2}\boldsymbol{W}^T\boldsymbol{W} + \frac{1}{2}C\sum_{i=1}^{m} e_i^2 - \sum_{i=1}^{m}\alpha_i\left(\boldsymbol{W}^T\varphi\left(\left[\boldsymbol{X}_h\boldsymbol{Y}_l^h\right]\right) + b + e_i - y_{hi}\right) \tag{4.54}$$

According to the Karush–Kuhn–Tucker conditions, by differentiating the above function with respect to the Lagrange multipliers and eliminating the variables w and e_i, the optimization problem can be transformed into a linear equation, as shown in Eq. (4.55).

$$\begin{bmatrix} 0 & \boldsymbol{e}^T \\ \boldsymbol{e} & \boldsymbol{K}+C^{-1}\boldsymbol{I} \end{bmatrix} \begin{bmatrix} b \\ \boldsymbol{\alpha} \end{bmatrix} = \begin{bmatrix} 0 \\ \boldsymbol{Y}_h \end{bmatrix} \tag{4.55}$$

where **K** is an m × m matrix with elements $K_{ij} = \varphi\left(\left[\boldsymbol{X}_{hi}\boldsymbol{Y}_{li}^h\right]\right)^T \varphi\left(\left[\boldsymbol{X}_{hj}\boldsymbol{Y}_{lj}^h\right]\right) = \mathrm{k}\left(\left[\boldsymbol{X}_{hi}\boldsymbol{Y}_{li}^h\right], \left[\boldsymbol{X}_{hj}\boldsymbol{Y}_{lj}^h\right]\right)$, with $\mathrm{k}\left(\left[\boldsymbol{X}_{hi}\boldsymbol{Y}_{li}^h\right], \left[\boldsymbol{X}_{hj}\boldsymbol{Y}_{lj}^h\right]\right)$ being the kernel function; $e^T = [1, \ldots, 1]_{1\times m}$ and $\alpha = [\alpha_1, \alpha_2, \ldots, \alpha_m]$. Thus, the resulting approximate PKI-LSSVR model can be expressed as follows:

$$\hat{F} = \sum_{i=1}^{m} \alpha_i k\left(\left[\boldsymbol{X}_{hi}\boldsymbol{Y}_{li}^h\right], \left[\boldsymbol{X}_{hj}\boldsymbol{Y}_{lj}^h\right]\right) + b \tag{4.56}$$

where α and b are solutions to Eq. (4.55).

Various kernel functions are available, including linear, polynomial, sigmoid and Gaussian kernels. Because of its advantages of relatively few parameters to set and excellent overall performance, the Gaussian kernel function is an effective and frequently used option (Keerthi and Lin 2003; Liao et al. 2011). This function is expressed as follows:

$$\mathrm{k}\left(\left[\boldsymbol{X}_{hi}\boldsymbol{Y}_{li}^h\right], \left[\boldsymbol{X}_{hj}\boldsymbol{Y}_{lj}^h\right]\right) = \exp\left(-\frac{\left[\boldsymbol{X}_{hi}\boldsymbol{Y}_{li}^h\right] - \left[\boldsymbol{X}_{hj}\boldsymbol{Y}_{lj}^h\right]^2}{2\sigma^2}\right) \tag{4.57}$$

Consequently, there are two hyper-parameters, the regularization parameter C and the kernel parameter r, which need to be appropriately chosen a priori in the PKI-LSSVR model to improve its generalizability.

The selection of hyper-parameters plays an important role in the performance of SVR. Finding the best combination of hyper-parameters is often a troublesome problem due to the high nonlinearity of the model performance with respect to these parameters. A popular approach for SVR tuning is to select the best choices among a certain set of candidate parameters by means of evolutionary methods (dos Santos et al. 2012), such as genetic algorithms (Wei and Zhang 2012), particle swarm optimization (Gilan et al. 2012) or ant colony optimization (Niu et al. 2010).

4.3 Co-Kriging Approaches

4.3.1 Traditional Co-Kriging Approach

Suppose that the HF model is $y_h : \mathbb{R}^m \to \mathbb{R}$ and that the LF model is $y_l : \mathbb{R}^m \to \mathbb{R}$. The sample sets are

$$S_l = (x_l^{(1)}, \ldots, x_l^{(n_l)})^T \in \mathbb{R}^{n_l \times m}$$
$$S_h = (x_h^{(1)}, \ldots, x_h^{(n_h)})^T \in \mathbb{R}^{n_h \times m} \tag{4.58}$$

The corresponding responses are

$$y_l = [y_l(x_l^{(1)}), \ldots, y_l(x_l^{(n_l)})]^T \in \mathbb{R}^{n_l}$$
$$y_h = [y_h(x_h^{(1)}), \ldots, y_h(x_h^{(n_h)})]^T \in \mathbb{R}^{n_h} \tag{4.59}$$

where n_l and n_h are the numbers of LF and HF sample points, respectively. Generally, it is assumed that $n_l \gg n_h$.

Then, the formula for Kennedy and O'Hagan's autoregressive model can be represented as

$$\hat{y}_h(x) = \rho \hat{y}_l(x) + \hat{y}_d(x) \tag{4.60}$$

where $\hat{y}_h(x)$ is the sum of two GP models, with ρ being a scaling factor. The hat symbols indicate that the models are approximations and $\hat{y}_d(x)$ represents a model of the discrepancy between the HF and LF models.

The applied distance measure between two sample points i and j is

$$d(\boldsymbol{x}^{(i)}, \boldsymbol{x}^{(j)}) = \sum_{k=1}^{m} \theta_k (\boldsymbol{x}_k^{(i)} - \boldsymbol{x}_k^{(j)})^{p_k} \tag{4.61}$$

where k is the number of dimensions, and θ_k and P_k are hyper-parameters tuned to the data at hand. The correlation function between points $x^{(i)}$ and $x^{(j)}$ is expressed as

$$\boldsymbol{R}(\boldsymbol{x}^{(i)}, \boldsymbol{x}^{(j)}) = \exp[-d(\boldsymbol{x}^{(i)}, \boldsymbol{x}^{(j)})] \tag{4.62}$$

When the response at a new point x is needed, the correlation vector $c(x)$ with the new point is formed, which is given below:

$$\boldsymbol{c}(\boldsymbol{x}) = \begin{bmatrix} \rho \sigma_l^2 \boldsymbol{R}_l(\boldsymbol{x}_l, \boldsymbol{x}) \\ \rho^2 \sigma_l^2 \boldsymbol{R}_l(\boldsymbol{x}_h, \boldsymbol{x}) + \sigma_d^2 \boldsymbol{R}_d(\boldsymbol{x}_h, \boldsymbol{x}) \end{bmatrix} \tag{4.63}$$

where σ^2 are the process variances, with the subscripts l and d indicating the parameters that belong to the LF and discrepancy models, respectively.

The prediction $\hat{y}_h(x)$ is calculated as follows:

$$\hat{y}_h(\boldsymbol{x}) = f(x)^T \beta^* + c(x)^T C^{-1}(y - F\beta^*) \tag{4.64}$$

where F and $f(x)$ are regression models that rely on the existing sampling data and the predicted points, respectively.

y, β^* and C can be expressed as

$$y = \begin{bmatrix} y_l \\ y_h \end{bmatrix}, \ \beta^* = (F^T C^{-1} F)^{-1} F^T \Lambda^{-1} y, \text{ and } \boldsymbol{C} = \begin{bmatrix} \sigma_l^2 \boldsymbol{R}_l(\boldsymbol{x}_l, \boldsymbol{x}_l) & \rho\sigma_l^2 \boldsymbol{R}_l(\boldsymbol{x}_l, \boldsymbol{x}_h) \\ \rho\sigma_l^2 \boldsymbol{R}_l(\boldsymbol{x}_l, \boldsymbol{x}_h) & \rho^2\sigma_l^2 \boldsymbol{R}_l(\boldsymbol{x}_h, \boldsymbol{x}_h) + \sigma_h^2 \boldsymbol{R}_h(\boldsymbol{x}_h, \boldsymbol{x}_h) \end{bmatrix} \tag{4.65}$$

The MSE of prediction is

$$\varphi(x) = c' + u^T (F^T C^{-1} F)^{-1} u - c^T C^{-1} c \tag{4.66}$$

where $u = F^T C^{-1} c - f$ and $c' = \rho^2 \sigma_l^2 + \sigma_d^2$.

4.3.2 *Extended Co-Kriging Approaches*

4.3.2.1 Hierarchical Kriging (HK) Approach

In this subsection, the hierarchical kriging (HK) method proposed by Han and Görtz (2012) is reviewed. In HK, the LF model is used to capture the general trend of the HF model, whereas the HF samples are used for calibration. The HK formulation can be expressed as

$$y_{MF}(\boldsymbol{x}) = y_m(\boldsymbol{x}) + Z(\boldsymbol{x}) \tag{4.67}$$

where $y_m(\boldsymbol{x}) = \beta_0 \hat{y}_{lf}(\boldsymbol{x})$ is the global trend function. Here, $\hat{y}_{lf}(\boldsymbol{x})$ is the LF surrogate model, which is built as a kriging surrogate model with the LF sample data. β_0 is a scaling factor indicating the influence of the LF model on the predictions of the HF model. $Z(\boldsymbol{x})$ is a stationary random process with zero mean and a covariance of

$$Cov[Z(\boldsymbol{x}), Z(\boldsymbol{x}')] = \sigma^2 R(\boldsymbol{x}, \boldsymbol{x}') \tag{4.68}$$

where σ is the process standard deviation determining the overall magnitude of the variance. $R(\boldsymbol{x}, \boldsymbol{x}') = \prod_{k=1}^{m} R_k(\theta_k, \boldsymbol{x}_k - \boldsymbol{x}'_k)$ is the spatial correlation function, which depends only on the Euclidean distance between two sites, $\boldsymbol{x}$ and $\boldsymbol{x}'$. $\boldsymbol{\theta} = (\theta_1, \ldots, \theta_m)$ is the unknown correlation vector to be determined. In this study, a Gaussian correlation function is adopted, which is given below:

$$R_k(\theta_k, \boldsymbol{x}_k - \boldsymbol{x}'_k) = \exp(-\theta_k (\boldsymbol{x}_k - \boldsymbol{x}'_k)^2) \tag{4.69}$$

Generally, $\boldsymbol{\theta}$ is obtained using maximum likelihood estimation. The likelihood function can be formulated as

$$L(\beta_0, \sigma^2, \boldsymbol{\theta}) = \frac{1}{\sqrt{(2\pi\sigma^2)^n |\boldsymbol{R}|}} \exp(-\frac{1}{2}\frac{(\boldsymbol{Y}_h - \beta_0\boldsymbol{F})^T \boldsymbol{R}^{-1}(\boldsymbol{Y}_h - \beta_0\boldsymbol{F})^T}{\sigma^2}) \tag{4.70}$$

where $\boldsymbol{Y}_h$ are the actual responses at the HF sample points and $\boldsymbol{F} = [\hat{y}_{lf}(\boldsymbol{x}_h^1) \ldots f(\hat{y}_{lf}\boldsymbol{x}_h^n)]^T$ is the vector of the values predicted by the LF surrogate model at the HF sample points $\boldsymbol{x}_h = (\boldsymbol{x}_h^1, \ldots, \boldsymbol{x}_h^n)$. The corresponding maximum likelihood estimates of the coefficient β_0 and the process variances σ^2 are expressed as (Han and Görtz 2012)

$$\beta_0(\boldsymbol{\theta}) = (\boldsymbol{F}^T\boldsymbol{R}^{-1}\boldsymbol{F})^{-1}\boldsymbol{F}^T\boldsymbol{R}^{-1}\boldsymbol{Y}_h \tag{4.71}$$

$$\sigma^2(\boldsymbol{\theta}, \beta_0) = \frac{1}{n}(\boldsymbol{Y}_h - \beta_0\boldsymbol{F})^T\boldsymbol{R}^{-1}(\boldsymbol{Y}_h - \beta_0\boldsymbol{F}) \tag{4.72}$$

where $\boldsymbol{R} := (R(\boldsymbol{x}_h^i, \boldsymbol{x}_h^j))_{i,j} \in R^{n\times n}$ is the correlation matrix and $\boldsymbol{r} := (\boldsymbol{R}(\boldsymbol{x}_h^i, \boldsymbol{x}))_i \in R^n$ is the correlation vector between an unobserved point $\boldsymbol{x}$ and the HF sample points.

Upon substituting Eqs. (4.71) and (4.72) into Eq. (4.70) and taking the logarithm, the following expression remains to be maximized:

$$\begin{aligned} &\max \phi(\Theta) = -n \ln \sigma^2(\boldsymbol{\theta}) - \ln|\boldsymbol{R}(\boldsymbol{\theta})| \\ &s.t.\ \Theta > 0 \end{aligned} \tag{4.73}$$

where Θ denotes the vector of $\boldsymbol{\theta}$, which is a function of both σ and R.

Finally, the HK predictor for unobserved points can be written as (Han and Görtz 2012)

$$y_{mf}(\boldsymbol{x}) = \beta_0\hat{y}_{lf}(\boldsymbol{x}) + \boldsymbol{r}^T(\boldsymbol{x})\boldsymbol{R}^{-1}(\boldsymbol{Y}_h - \beta_0\boldsymbol{F}) \tag{4.74}$$

4.3.2.2 An Improved HK Approach

In HK, the scaling factor β_0 is a constant, which indicates that the proportional relation between the HF and LF models is the same throughout the whole design space. However, in reality, the specific relationship between the HF and LF models is unknown and may vary at different points. Therefore, using a constant to express this relationship is not sufficiently accurate. Furthermore, β_0 is calculated based only on the initial sample points, which are highly correlated with the method used to select them. In fact, when generating initial sample points, designers generally attempt to ensure that their distribution will be as homogeneous as possible to cover a wider area; consequently, however, the areas close to the global or local maximum/minimum may not be covered, meaning that HK is unable to provide a reasonable prediction in these areas. Based on these considerations, an improved hierarchical kriging (IHK) model is proposed to address this problem.

The IHK formulation can be represented as follows:

$$Y(x) = h(x)\hat{y}_{lf}(x) + z(x) \tag{4.75}$$

Here, $h(x)$ is a response surface model with the following expression:

$$h(x) = \beta_0 + \sum_{i=1}^{n} \beta_i x_i + \sum_{1 \le j \le k \le n} \beta_{jk} x_j x_k \tag{4.76}$$

where n denotes the dimensionality of the design space; x_i, x_j, x_k are different design variables and β_i, β_{jk} are the coefficients in front of them. The other parameters are the same as in HK. In IHK, the value of the global trend function $h(x)$ varies with the predictor and thus can express the relationship between the HF and LF functions more specifically; consequently, IHK shows better global performance than HK does. Below, the solution procedure for IHK will be demonstrated in a 2D case; the same procedure is also applicable in higher dimensional cases.

Model solving

The IHK model has the following expression in the 2D case:

$$\hat{y}(x_1, x_2) = (\beta_1 + \beta_2 x_1 + \beta_3 x_2 + \beta_4 x_1^2 + \beta_5 x_1 x_2 + \beta_6 x_2^2)\hat{y}_{lf}(x_1, x_2) + z(x_1, x_2) \tag{4.77}$$

It can also be written as

$$y(x) = f(x)^T \beta + z(x) \tag{4.78}$$

where

$$\beta = [\beta_1 \; \beta_2 \ldots \beta_6]$$

$$\begin{aligned} f(x) &= [f_1 \; f_2 f_3 f_4 f_5 f_6] \\ &= [\hat{y}_{lf}(x_1, x_2) \; x_1 \hat{y}_{lf}(x_1, x_2) \; x_2 \hat{y}_{lf}(x_1, x_2) \; x_1^2 \hat{y}_{lf}(x_1, x_2) \; x_1 x_2 \hat{y}_{lf}(x_1, x_2) \; x_2^2 \hat{y}_{lf}(x_1, x_2)] \end{aligned}$$

Higher dimensional problems can also be written in the same form as that of Eq. (4.77). The expression for IHK is similar to that for universal kriging; therefore, the solution method for universal kriging can also be used here. Under the assumption that the response of the HF model can be approximated as a linear combination of the chosen HF data y_s, the IHK predictor can be written as

$$\hat{y}(x) = w^T y_s \tag{4.79}$$

where $w = [w^{(1)}, \ldots, w^{(n)}]$ is a vector of weight coefficients associated with the sampled HF data. Then, by replacing $y_s = [y^{(1)}, \ldots, y^{(n)}]^T$ with random quantities

$Y_s = [Y^{(1)}, \ldots, Y^{(n)}]^T$, the error between the predicted and true responses can be written as

$$\begin{aligned}\hat{y}(x) - y(x) &= w^T Y - y(x) \\ &= w^T (F\beta + Z) - (f(x)^T \beta + z) \\ &= w^T Z - z + (F^T w - f(x))^T \beta\end{aligned} \tag{4.80}$$

For the predictor to remain unbiased, it is necessary that $F^T w - f(x) = 0$; thus,

$$F^T w(x) = f(x) \tag{4.81}$$

Subject to the constraint given in Eq. (4.81), the MSE of the predictor expressed in Eq. (4.79), which is to be minimized, is

$$\begin{aligned}\varphi(x) &= E[(\hat{y}(x) - y(x))^2] \\ &= E[(w^T Z - z)^2] \\ &= [z^2 + w^T Z Z^T w - 2w^T Z z] \\ &= \sigma^2(1 + \mathrm{w}^{\mathrm{T}} R w - 2w^T r)\end{aligned} \tag{4.82}$$

For the constrained problem, the Lagrange multiplier λ can be introduced, and the Lagrangian function for the problem of minimizing φ with respect to ω is

$$L(w, \lambda) = \sigma^2(1 + w^T R w - 2w^T r) - \lambda^T (F^T w - f) \tag{4.83}$$

The gradient of Eq. (4.83) with respect to w is

$$L_c'(w, \lambda) = 2\sigma^2(Rw - r) - F\lambda \tag{4.84}$$

From the first-order necessary conditions for optimality, the following system of equations can be obtained:

$$\begin{bmatrix} R & F \\ F^T & 0 \end{bmatrix} \begin{bmatrix} w \\ \tilde{\lambda} \end{bmatrix} = \begin{bmatrix} r \\ f \end{bmatrix} \tag{4.85}$$

where

$F = [f(x^{(1)}) \ldots f(x^{(n)})]^T$, $\tilde{\lambda} = -\frac{\lambda}{2\sigma^2}$

$R := (R(x^{(i)}, x^{(j)}))_{i,j} \in R^{n \times n}$, $r := (R(x^{(i)}, x))_i \in R^n$

The matrix w can be obtained by solving the above equations, and the solution to Eq. (4.85) is as follows:

$$\begin{aligned}\tilde{\lambda} &= (F^T R^{-1} F)^{-1}(F^T R^{-1} r - f) \\ w &= R^{-1}(r - F\tilde{\lambda})\end{aligned} \tag{4.86}$$

Hence, the IHK predictor at any unobserved x is found to be

$$\begin{aligned}\hat{y}(x) &= f^T \beta^* + r^T R^{-1}(Y - F\beta^*) \\ &= f(x)^T \beta^* + r(x)^T \gamma^*\end{aligned} \tag{4.87}$$

where $\beta^* = (F^T R^{-1} F)^{-1} F^T R^{-1} Y$ is the coefficient of the scaling factor. Note that for a fixed set of design data, the matrices β^* and γ^* are fixed, and they can be calculated as part of the model-fitting process in this method. For every new x, only the vectors $f(x)$ and $r(x)$ need to be computed, adding two simple products. Because a kriging model is an interpolative Bayesian surrogate model, the MSE of the model will be zero at all sample points. The expression for the MSE at an unobserved point can be written as

$$\varphi(x) = \sigma^2[1 + (F^T R^{-1} r - f)^T (F^T R^{-1} F)^{-1} (F^T R^{-1} r - f) - r^T R^{-1} r] \tag{4.88}$$

Although the MSE estimation for IHK highly resembles that for HK, the matrix f here is related not only to the response of the LF model but also to the location of the predictor. Thus, the behaviour of the LF model is also explicitly tuned here, thus permitting a better MSE estimation than in the HK case.

Correlation model

The correlation function must be calculated during the model-building stage, and it often has the following form:

$$R(x, x') = \prod_{j=1}^{m} R_j(\theta, x_j - x'_j) \tag{4.89}$$

where $\theta = (\theta_1, \ldots, \theta_m) \in R^m$ are the hyper-parameters to be tuned and m denotes the dimensionality of the design space. The most popular form of this correlation function is a Gaussian exponential function, which can be calculated as follows:

$$\begin{aligned}R_k(\theta_k, x_k - x'_k) &= \exp(-\theta_k |x_k - x'_k|^{p_k}) \\ R(\theta, x, x') &= \prod_{k=1}^{m} \exp(-\theta_k |x_k - x'_k|^{p_k}),\ 1 \le p_k \le 2\end{aligned} \tag{4.90}$$

A cubic spline correlation function (Lophaven et al. 2002c) is also often used because it is always second-order differentiable, which is an expected characteristic when combined with gradient information. This function is given by

$$R_k(\theta_k, x_k - x'_k) = \begin{cases} 1 - 15\xi_k^2 + 30\xi_k^3 & for\, 0 \le \xi_k \le 0.2 \\ 1.25(1 - \xi_k)^3 & for\ 0.2 < \xi_k < 1, \\ 0 & for\ \xi_k \ge 1 \end{cases} \quad (4.91)$$
$$where\ \xi_k = \theta_k |x_k - x'_k|$$

Hyper-parameter tuning strategy

The $\theta = (\theta_1, \ldots, \theta_m)$ in Eq. (4.89) affect the attenuation rate of the correlation function, with a larger θ leading to a faster decrease. Generally, the unknown parameters θ are found using maximum likelihood estimation. The likelihood function can be formulated as follows:

$$L(\beta, \sigma^2, \theta) = \frac{1}{\sqrt{(2\pi\sigma^2)^n |R|}} \exp(-\frac{1}{2} \frac{(Y - F\beta^*)^T R^{-1} (Y - F\beta^*)^T}{\sigma^2}) \quad (4.92)$$

The corresponding maximum likelihood estimates of the coefficient β and the process variances are

$$\begin{aligned} \beta* &= (F^T R^{-1} F)^{-1} F^T R^{-1} Y \\ \sigma^2 &= \frac{1}{m} (Y - F\beta^*)^T R^{-1} (Y - F\beta^*) \end{aligned} \quad (4.93)$$

Upon substituting Eq. (4.93) into Eq. (4.92) and taking the logarithm, the following expression remains to be maximized:

$$\begin{aligned} \max \phi(\Theta) = -n \ln \sigma^2(\theta) - \ln|R(\theta)| \\ s.t.\ \Theta > 0 \end{aligned} \quad (4.94)$$

where Θ denotes the vector of θ and both σ and R are functions of Θ. Following the method proposed in Lophaven et al. (2002b), this problem can be solved by using a modified version of the direct search algorithm of Hooke and Jeeves.

References

Aute V, Saleh K, Abdelaziz O, Azarm S, Radermacher R (2013) Cross-validation based single response adaptive design of experiments for Kriging metamodeling of deterministic computer simulations. Struct Multidiscip Optim 48:581–605

Bandler JW, Cheng QS, Nikolova NK, Ismail MA (2004) Implicit space mapping optimization exploiting preassigned parameters. IEEE Trans Microw Theory Tech 52:378–385

Burgee S, Giunta AA, Balabanov V, Grossman B, Mason WH, Narducci R, Haftka RT, Watson LT (1996) A coarse-grained parallel variable-complexity multidisciplinary optimization paradigm. Int J Supercomput Appl High Perform Comput 10:269–299

Chen S, Jiang Z, Yang S, Apley DW, Chen W (2016) Nonhierarchical multi-model fusion using spatial random processes. Int J Numer Meth Eng 106:503–526
Cheng GH, Younis A, Hajikolaei KH, Wang GG (2015) Trust region based mode pursuing sampling method for global optimization of high dimensional design problems. J Mech Des 137:021407
dos Santos GS, Luvizotto LGJ, Mariani VC, dos Santos Coelho L (2012) Least squares support vector machines with tuning based on chaotic differential evolution approach applied to the identification of a thermal process. Expert Syst Appl 39:4805–4812
Forrester AI, Sóbester A, Keane AJ (2007) Multi-fidelity optimization via surrogate modelling. Proc R Soc A Math Phys Eng Sci 463:3251–3269
Gano SE, Renaud JE, Sanders B (2005) Hybrid variable fidelity optimization by using a Kriging-based scaling function. AIAA J 43:2422–2433
Gano SE, Renaud JE, Martin JD, Simpson TW (2006) Update strategies for kriging models used in variable fidelity optimization. Struct Multidiscip Optim 32:287–298
Gilan SS, Jovein HB, Ramezanianpour AA (2012) Hybrid support vector regression–Particle swarm optimization for prediction of compressive strength and RCPT of concretes containing metakaolin. Constr Build Mater 34:321–329
Haftka RT (1991) Combining global and local approximations. AIAA Journal 29:1523–1525
Han Z-H, Görtz S (2012) Hierarchical Kriging model for variable-fidelity surrogate modeling. AIAA J 50:1885–1896
Han Z, Zimmerman R, Görtz S (2012) Alternative cokriging method for variable-fidelity surrogate modeling. AIAA J 50:1205–1210
Hu J, Zhou Q, Jiang P, Shao X, Xie T (2017) An adaptive sampling method for variable-fidelity surrogate models using improved hierarchical kriging. Eng Optim 1–19
Huang D, Allen TT, Notz WI, Miller RA (2006) Sequential kriging optimization using multiple-fidelity evaluations. Struct Multidiscip Optim 32:369–382
Huntington D, Lyrintzis C (1998) Improvements to and limitations of Latin hypercube sampling. Probab Eng Mech 13:245–253
Jin R, Chen W, Sudjianto A (2005) An efficient algorithm for constructing optimal design of computer experiments. J Stat Plan Inference 134:268–287
Journel AG, Huijbregts CJ (1978) Mining geostatistics. Academic Press, London
Keerthi SS, Lin CJ (2003) Asymptotic behaviors of support vector machines with Gaussian kernel. Neural Comput 15:1667–1689
Kennedy MC, O'Hagan A (2000) Predicting the output from a complex computer code when fast approximations are available. Biometrika 87:1–13
Kleijnen JP (2008) Design and analysis of simulation experiments. Springer
Kleijnen JPC (2009) Kriging metamodeling in simulation: a review. Eur J Oper Res 192:707–716
Kleijnen JPC (2017) Regression and Kriging metamodels with their experimental designs in simulation: a review. Eur J Oper Res 256:1–16
Le Gratiet L, Garnier J (2014) Recursive co-kriging model for design of computer experiments with multiple levels of fidelity. Int J Uncertain Quantif 4
Lewis R, Nash, S (2000) A multigrid approach to the optimization of systems governed by differential equations. In: 8th symposium on multidisciplinary analysis and optimization, p 4890
Liao R, Zheng H, Grzybowski S, Yang L (2011) Particle swarm optimization-least squares support vector regression based forecasting model on dissolved gases in oil-filled power transformers. Electr Power Syst Res 81:2074–2080
Liu Y, Collette M (2014) Improving surrogate-assisted variable fidelity multi-objective optimization using a clustering algorithm. Appl Soft Comput 24:482–493
Lophaven S, Nielsen H, Sondergaard J (2002a) DACE: a Matlab Kriging toolbox, version 2.0. IMM Technical University of Denmark, Lyngby, Denmark
Lophaven SN, Nielsen HB, Søndergaard J (2002b) Aspects of the matlab toolbox DACE. Informatics and mathematical modelling, Technical University of Denmark, DTU
Lophaven SN, Nielsen HB, Søndergaard J (2002c) DACE-A Matlab Kriging toolbox, version 2.0

Martin JD, Simpson TW (2004) A Monte Carlo simulation of the Kriging model. In: 10th AIAA/ISSMO symposium on multidisciplinary analysis and optimization, pp 2004–4483

Martin JD, Simpson TW (2005) Use of kriging models to approximate deterministic computer models. AIAA J 43:853–863

Niu D, Wang Y, Wu DD (2010) Power load forecasting using support vector machine and ant colony optimization. Expert Syst Appl 37:2531–2539

Ollar J, Mortished C, Jones R, Sienz J, Toropov V (2017) Gradient based hyper-parameter optimisation for well conditioned kriging metamodels. Struct Multidiscip Optim 55:2029–2044

Park J-S (1994) Optimal Latin-hypercube designs for computer experiments. J Stat Plan Inference 39:95–111

Rayas-Sanchez JE (2016) Power in simplicity with ASM: tracing the aggressive space mapping algorithm over two decades of development and engineering applications. IEEE Microw Mag 17:64–76

Sun G, Li G, Li Q (2012) Variable fidelity design based surrogate and artificial bee colony algorithm for sheet metal forming process. Finite Elem Anal Des 59:76–90

Tyan M, Nguyen NV, Lee J-W (2015) Improving variable-fidelity modelling by exploring global design space and radial basis function networks for aerofoil design. Eng Optim 47:885–908

Van Nguyen N, Tyan M, Lee J-W (2015) A modified variable complexity modeling for efficient multidisciplinary aircraft conceptual design. Optim Eng 16:483–505

Viana FA, Steffen V, Butkewitsch S, de Freitas Leal M (2009) Optimization of aircraft structural components by using nature-inspired algorithms and multi-fidelity approximations. J Global Optim 45:427–449

Viana FA, Simpson TW, Balabanov V, Toropov V (2014) Special section on multidisciplinary design optimization: metamodeling in multidisciplinary design optimization: how far have we really come? AIAA J 52:670–690

Wei SF, Zhang Y (2012) Application of least squares support vector regression trained by genetic algorithm in reliability prediction of LAN/WLAN integration network. In: Advanced Materials Research, vol. 433. Trans Tech Publications

Xiong S, Qian PZ, Wu CJ (2013) Sequential design and analysis of high-accuracy and low-accuracy computer codes. Technometrics 55:37–46

Zheng J, Shao X, Gao L, Jiang P, Li Z (2013) A hybrid variable-fidelity global approximation modelling method combining tuned radial basis function base and kriging correction. J Eng Des 24:604–622

Zheng J, Shao X, Gao L, Jiang P, Qiu H (2014) A prior-knowledge input LSSVR metamodeling method with tuning based on cellular particle swarm optimization for engineering design. Expert Syst Appl 41:2111–2125

Zhou X, Jiang T (2016) Metamodel selection based on stepwise regression. Struct Multidiscip Optim 1–17

Zhou Q, Shao X, Jiang P, Zhou H, Shu L (2015) An adaptive global variable fidelity metamodeling strategy using a support vector regression based scaling function. Simul Model Pract Theory 59:18–35

Zhou Q, Shao X, Jiang P, Gao Z, Wang C, Shu L (2016a) An active learning metamodeling approach by sequentially exploiting difference information from variable-fidelity models. Adv Eng Inform 30:283–297

Zhou Q, Shao X, Jiang P, Gao Z, Zhou H, Shu L (2016b) An active learning variable-fidelity metamodelling approach based on ensemble of metamodels and objective-oriented sequential sampling. J Eng Des 27:205–231

Zhou Q, Jiang P, Shao X, Hu J, Cao L, Wan L (2017) A variable fidelity information fusion method based on radial basis function. Adv Eng Inform 32:26–39

Zimmermann R, Han Z-H (2010) Simplified cross-correlation estimation for multi-fidelity surrogate cokriging models. Adv Appl Math Sci 7:181–202

Chapter 5
Verification Methods for Surrogate Models

A surrogate model built based on a limited number of sample points will inevitably have large prediction uncertainty. Applying such imprecise surrogate models in design and optimization may lead to misleading predictions or optimal solutions located in unfeasible regions (Picheny 2009). Therefore, verifying the accuracy of a surrogate model before using it can ensure the reliability of the design. The following steps are often used to verify the accuracy of a surrogate model:

Step 1: Predict the response $\hat{y}$ at an unobserved point. This response is the output of the surrogate model; sometimes, it can also be the outcome of a simulation model, mathematical formula, etc.
Step 2: Obtain the true observation y at the unobserved point. The true response y is usually the outcome of a simulation model or physical experiment.
Step 3: Compare the predicted response $\hat{y}$ and the true observation y. The comparison between the two responses will reveal how close they are. Various error methods can be used in this step, and the errors at different points can be considered collectively to obtain an overall assessment of the model. Sometimes, qualitative judgements can also be applied, such as graphic comparisons or hypothesis testing.

The applications of error metrics in the design optimization can be generally classified into four cases (Acar 2015): (1) identifying regions with relatively high uncertainty in the input domain to determine promising areas for model refinement; (2) obtaining an overall assessment of the constructed surrogate model to be used for prediction, uncertainty quantification or optimization; (3) for an ensemble of surrogate models, determining the optimal weight factors for the individual surrogate models; and (4) choosing the most appropriate model among the alternatives when multiple surrogate models are available. In cases (1), (3) and (4), after the application of error metrics for these three purposes, the general level of uncertainty of the surrogate model should be known, but the metric values do not represent the true error. That is, it is not necessary to evaluate the true fidelity of the surrogate model compared with the high-fidelity (HF) simulation model or experimental model, since evaluating the true error of a surrogate model is sometimes

P. Jiang et al., *Surrogate Model-Based Engineering Design and Optimization*,
Springer Tracts in Mechanical Engineering,
https://doi.org/10.1007/978-981-15-0731-1_5

time-consuming and may require additional validation points. For example, in case (1), only regions with relatively high uncertainty need to be detected with the error metric, while in case (4), comparing the degree of uncertainty of one surrogate model over the others among the alternatives is sufficient for the design task. However, in case (2), in which the designers must decide whether to accept or reject the constructed surrogate model, the overall accuracy of the surrogate model needs to be known, and the metric value should be as close to the true error as possible since the criterion for accepting a model is often set in reference to the true error.

There are many alternatives available when choosing an error metric to assess the accuracy of a surrogate model, such as the mean square error (MSE), the mean absolute error (MeanAE), the leave-one-out (LOO) error, the bootstrap error (BE), the predictive estimation of model fidelity (PEMF) error and the root mean square error (RMSE). There are many different possible criteria for classifying the existing error metrics. For example, based on whether an additional verification sample set is needed, these metrics can be divided into metrics that rely on testing methods and metrics that rely on sampling methods. The former metrics evaluate the accuracy of a surrogate model based on the same sample sets used for modelling, and the evaluation results are highly dependent on the error metric itself. The latter metrics require additional test samples, and their performance is mainly influenced by the method of selecting the validation test. Moreover, metrics that rely on sampling methods may not be affordable when the simulation cost is high. Based on the region of the design domain in which the verification method works, the existing methods can also be classified into methods that quantify the global error and methods that quantify local errors (Goel et al. 2007). Methods that quantify the global error generally include three different types (Queipo et al. 2005): split sampling, bootstrapping and the Akaike's information criterion (AIC) method. Methods for measuring local errors include (1) the prediction MSE for kriging and (2) the linear reference model (LRM) (Nguyen et al. 2011). Based on whether the error metric depends on the specific model, the existing error metrics can additionally be roughly classified as either parametric (model-based) or distribution-free (model-independent) metrics (Goel et al. 2009). Parametric error metrics are generally based on statistical assumptions. One example is the prediction variance (PV) of a polynomial response surface (PRS) model, which is computed based on the assumptions that the predictions contain normally distributed noise of zero mean and that the variance is the same everywhere and uncorrelated. If these assumptions are invalid, the predictions may exhibit a large error. Model-independent error metrics can be applied to more complex problems, but they may incur higher computational costs. Distribution-free verification methods are not limited to any one surrogate model or specific kinds of surrogate models, and they generally have wider applications. For other classifications of verification models for surrogate models, readers can refer to Hyndman and Koehler (2006), Li and Heap (2011).

In this chapter, a number of error metrics are first introduced, separated into two different classes based on whether an additional verification sample set is needed. Then, a review of the applications of these error metrics in engineering design is presented. The performances of the metrics are investigated for problems with various characteristics. Finally, some guidelines for selecting suitable verification methods are provided.

5.1 Metrics Relying on Testing Methods

5.1.1 Jackknife Error

Quenouille (1949) first introduced the jackknife method, in which the bias of a serial correlation estimator is estimated by removing part of the original sample set, recalculating the estimator based on the remaining samples and comparing the differences between the predicted responses. Let $x_1, x_2, \ldots, x_n$ denote the set of samples, which are randomly, independently and identically distributed. $H_n = \hat{f}_n(x_1, x_2, \ldots, x_n)$ is the estimator of a parameter θ based on this sample set of size n. The bias of this estimator can be defined as (Miller 1974)

$$bias(H_n) = \hat{f}_n(x_1, x_2, \ldots, x_n) - \theta \tag{5.1}$$

When sample x_k is removed, the estimator for sample x_k becomes $H_{n-1,k} = \hat{f}_{n-1}(x_1, \ldots, x_{k-1}, x_{k+1}, \ldots, x_n)$. If this process is repeated for $k = 1, 2, \ldots, n$ until each sample has been deleted only once, n jackknife estimators, one for each sample, will be obtained. The jackknife predictor of the standard error can then be expressed as (Efron and Tibshirani 1993a):

$$\hat{S}_{JACK} = \sqrt{\frac{n-1}{n}\sum_{k=1}^{n}(H_{n-1,k} - \bar{H}_n)^2} \tag{5.2}$$

where $\bar{H}_n = \frac{1}{n}\sum_{i=1}^{n} H_{n-1,i}$. The factor $(n-1)/n$ is selected to ensure that $\hat{S}^2_{JACK}$ is an unbiased variance estimator of the sample mean value. As the size of the sample set increases, the delete-1 jackknife error may begin to incur an excessive computational cost; in this case, a grouped jackknife computation can be used. In the grouped jackknife approach, the original sample set of size n is divided into p groups of equal size $(n = pg)$, and each time, a randomly selected group is deleted instead of only one point. After the procedure has been repeated p times, such that each group has been deleted once and only once, the grouped jackknife predictor of the standard error is given as

$$\hat{S}_{JACK,p} = \sqrt{\frac{p-1}{p}\sum_{k=1}^{p}(H_{p-1,k} - \bar{H}_p)^2} \tag{5.3}$$

where $\bar{H}_p = \frac{1}{p}\sum_{i=1}^{p} H_{p-1,i}$ is the same as before. The jackknife error is a distribution-free error metric and does not require the theoretical formulas that are needed in traditional approaches (Shao and Tu 2012).

5.1.2 Bootstrap Error (BE)

The bootstrap approach was first proposed in 1993 (Efron and Tibshirani 1993b). There are two types of bootstrap methods: the nonparametric bootstrap method and the parametric bootstrap method. In the parametric bootstrap method, the original data are fitted to a certain type of distribution to estimate the shape parameters, and sampling is then performed from the obtained distribution using the Monte Carlo sampling method. For the nonparametric bootstrap method, there is no need to predefine the form of the distribution, and thus, this method is more widely used. Bootstrap observations are generally obtained through resampling with replacement from the original sample set m times. Resampling with replacement means that for a sample set of size n, the probability that each sample may be chosen each time is $1/n$. With m sampling iterations, this sampling method may result in one original sample, say x_i, being selected all m times, while the other $n-1$ samples are never sampled. Of course, the probability of this event occurring is very small, but it is still possible for it to happen. Let the frequencies with which the samples $x_1, x_2, \ldots, x_n$ occur in the bootstrapped sample set be denoted by $f_1, f_2, \ldots, f_n$, respectively; these frequencies should satisfy $f_1 + f_2 + \cdots + f_n = m$ and should follow a multinomial distribution.

When the bootstrap method is used to verify the accuracy of a surrogate model, through the m iterations of resampling with replacement, m bootstrapped subsets will be obtained. For each subset, the bootstrapped samples are the training set, and the remaining samples from the original sample set are the test set. The training set is used to construct an intermediate surrogate model, while the test set is utilized to validate its accuracy. Given a collection of m subsets $\{S_1, S_2, \ldots, S_m\}$, the LOO BE can be defined as

$$e_{boot} = \frac{1}{m}\sum_{i=1}^{m} err_{-i} \tag{5.4}$$

This estimator is biased upwards, but the estimation error is considered to show less variance compared with the cross-validation error. It has been found that when this metric is applied for model selection, the bootstrap selection procedure is not consistent, but the consistency can be improved with appropriate modifications (Shao 1996).

Since the subsets are sampled with replacement, the probability that any given sample point will not be chosen in each subset is $(1 - 1/n)^n \approx e^{-1} \approx 0.368$; the expected number of samples from the original data set appearing in the test set is thus $0.632n$. The 0.632 bootstrap estimation method was proposed by Efron (1983), and the corresponding metric can be formulated as follows:

$$e_{boot}^{0.632} = \frac{1}{n}\sum_{i=1}^{n}(0.632 \cdot \varepsilon 0_i + 0.368 \cdot acc_s) \tag{5.5}$$

where $\varepsilon 0_i$ is the error estimate for bootstrap sample i, and acc_s is the re-substitution accuracy estimate on the full data set (i.e. the accuracy on the training set). The upward bias of the original BE is corrected by averaging it with a downwardly biased estimator in the $e_{boot}^{0.632}$ calculation. In some interpolation models, the re-substitution error is 0 in Eq. (5.5). Therefore, the 0.632 BE can be written as

$$e_{boot}^{0.632} = \frac{1}{n}\sum_{i=1}^{n}0.632 \cdot \varepsilon 0_i \tag{5.6}$$

However, in situations in which the model is severely overfitted, the estimator $e_{boot}^{0.632}$ is downwardly biased because $acc_s = 0$ in these cases. To overcome this drawback, Efron and Tibshirani (1997) proposed the 0.632 + bootstrap metric, in which a greater weight is assigned to the acc_s term when the overfitting is larger to obtain a less-biased compromise between the two terms in Eq. (5.5). The algorithm for calculating $e_{boot}^{0.632+}$ is presented as Algorithm 5.1.

Algorithm 5.1: Calculating the $e_{boot}^{0.632+}$ error

1. Calculate the no-information error θ. θ can be estimated as the loss of all N_{test}^2 sample pairs (y_i, x_j):
$\hat{\theta} = \frac{1}{N_{test}^2}\sum_{i,j} L(y_i, \hat{f}(x_i)) \quad (i,j = 1, 2, \ldots, N_{test})$

2. Calculate the relative overfitting rate:
$\hat{R} = (\varepsilon 0_i - acc_s)/(\hat{\theta} - acc_s)$

3. The $e_{boot}^{0.632+}$ BE estimator is given by
$e_{boot}^{0.632+} = \hat{w} \cdot \varepsilon 0_i + (1 - \hat{\theta}) \cdot acc_s$
where $\hat{w} = 0.632/(1 - 0.368 \cdot \hat{\theta})$. The range of variation of the weight $\hat{w}$ is from 0.632 (no overfitting) to 1 (severe overfitting)

It is important to note that when the bootstrap method is used, repeated sample points may be obtained during the resampling process, which may lead to the failure of kriging model construction. Therefore, a small amount of random noise is sometimes added to the training points (Efron 1979).

5.1.3 Cross-Validation Error

Cross-validation (CV) was originally proposed in the 1930s (Larson 1931), and it was further developed and refined in the 1970s by Stone (Stone 1974; Salkind 2010). In the CV method, the uncertainty of a surrogate model is evaluated by splitting the current sampling set into training data and test data. In k-fold CV, the n sample points are divided into k disjoint subsets of equal size $m = n/k$. The model is trained k times, each time using all of the subsets t_h $(h = 1, 2, \ldots, k)$ except one for model training, and the remaining subset is used to calculate the prediction error. The k-fold CV estimator is defined as the mean of the k errors calculated in this way:

$$e^{CV} = \frac{1}{k}\sum_{i=1}^{k}\frac{1}{m}\sum_{j\in t_h} F(y_j, \hat{f}_{-i}(x_j)) \tag{5.7}$$

where $\hat{f}_{-i}(x_j)$ is the value at sample x_j predicted by the surrogate model built based on all subsets except subset i.

The k-fold CV method is a biased estimation method; however, the bias can be reduced by increasing the number of folds k (Kohavi 1995). Meckesheimer et al. (2002) investigated the influence of k for different kinds of surrogate models and suggested the use of $k = 1$ for low-order polynomial as well as radial basis function (RBF) models and the use of $k = 0.1n$ or $\sqrt{n}$ for kriging models, where n is the number of sample points used to construct the surrogate model. Rodriguez et al. (2010) also studied the changes in the bias and variance with different k values and found that $k = 2$ results in the most strongly biased estimator. In their work, $k = 5$ or $k = 10$ was recommended for use in error estimation to achieve a good balance between bias and computational cost.

Leave-one-out cross-validation (LOO-CV) is a special case of k-fold CV (Borra and Di Ciaccio 2010). In LOO-CV, k is equal to the number of sample points n; thus, each time, only one point is left out, and the other $n - 1$ points are utilized to build the intermediate surrogate model. During the LOO validation process, n intermediate surrogate models are built. The LOO error at point x_i is the difference between the response $\hat{y}_{-i}$ predicted by the i-th surrogate and the true response y_i. The sum of the LOO errors at all points is referred to as the generalized cross-validation (GCV) error, which can be formulated as follows:

$$e_{GCV} = \sum_{i=1}^{k} |\hat{y}_{-i}(x_i) - y_i(x_i)| \tag{5.8}$$

Based on the GCV error, the predicted residual error sum of squares (PRESS) measure is widely used when applying the LOO-CV method to assess the accuracy of a surrogate model (Vehtari et al. 2017). The PRESS is formulated as follows:

$$e_{PRESS} = \frac{1}{k}\sum_{i=1}^{k}\left|\hat{y}_{-i}(x_i) - y_i(x_i)\right| \tag{5.9}$$

The root mean square of the PRESS is calculated as follows:

$$\sigma_{PRESS} = \sqrt{e_{PRESS}/k} \tag{5.10}$$

The main disadvantage of k-fold CV is that the k training sets are not independent of each other, i.e. different training sets and/or test sets may contain some of the same data. This may lead to a large variance of the CV estimator (Breiman 1996). Several researchers have attempted to estimate this variance, but it has been shown that no biased estimator of the variance can be calculated. The LOO-CV method can provide a nearly unbiased estimate of the generalization error of a model. In this verification method, each point will appear in the training set $k - 1$ times and will appear in the test set only once. However, compared with that of the k-fold CV error ($k > 1$), the variance may be increased. For a sufficiently small sample set, the variance of the estimation may be unacceptable. To alleviate this problem, some researchers have defined certain bounds to ensure that the performance of LOO-CV will never be worse than that of the apparent error estimator under weak assumptions of error stability. In addition, some bias-corrected versions of the CV error have been proposed by Burman (1989) and Yanagihara et al. (2006). In the first study, bias correction of the k-fold CV error was considered, whereas the latter addressed only the bias correction of the LOO-CV error. Fushiki (2011) defined two families that relate the k-fold CV error to the training error and tested the performance of Burman's method on several test problems. Another problem with the LOO-CV error is that its value may be sensitive to a few large individual errors. Therefore, the mean pointwise cross-validation error ratio (PRESS-ratio) has been proposed to alleviate this problem (Goel and Stander 2009). The formula for the PRESS-ratio is given as follows:

$$\sigma_{ratio} = \frac{1}{k}\sum_{i=1}^{k}\left|\frac{\hat{y}_{-i}(x_i) - y_i(x_i)}{\hat{y}_{-i}(x_i)}\right| \tag{5.11}$$

The LOO error at each point is scaled by the predicted response at that point; thus, the magnitudes of the errors at all points are adjusted to a comparable level, and more importance can be placed on smaller values.

5.1.4 Prediction Variance (PV)

For some specific kinds of surrogate models such as kriging models, when such a model is used to predict the responses at unobserved points, the PV can also be easily calculated at the same time. The PV can be used as a metric for assessing the accuracy of surrogate models. The PV is a local error measurement metric, with a larger PV at a certain point indicating a larger range of uncertainty of the prediction at that point. In a kriging model, the prediction errors at two different points are not independent of each other, and the correlation between them is usually related to the distance between them. The larger the distance is, the weaker the correlation. Thus, for a point that is close to one or more sample points, the prediction error will tend to be smaller; the error will be very high at a point that is far away from all samples; and if a point overlaps with any point in the sample set, the prediction error at that point will be zero. The derivation and formulas for the PV of a kriging model can be found in Goel et al. (2009), Liu et al. (2012). For the example shown in Fig. 5.1a, the red line is the kriging approximation, the black line is the true function, and the pentagrams are the sample points. The blue shadowed area corresponds to the prediction uncertainty and represents the estimated area in which the responses may lie. The prediction MSE at each point and the corresponding true errors are plotted in Fig. 5.1b. Although there is some difference between the two curves, the prediction MSE still reflects the variation tendency of the true error as well as the rough location of the maximum error. This metric provides a rapid alternative for assessing the model quality and can be useful for model selection or model refinement.

The average MSE over a set of points can be used as the error metric to reflect the overall uncertainty of a surrogate model, as shown in Eq. (5.12).

$$e_{mse} = \frac{1}{n}\sum_{k=1}^{n}\sqrt{\hat{\sigma}^2(x_k)} \tag{5.12}$$

where $X = \{x_1, x_2, \ldots, x_n\}$ represents a randomly generated set of unobserved points with n elements, and $\hat{\sigma}^2(x_k)$ denotes the PV at unobserved point x_k. A lower

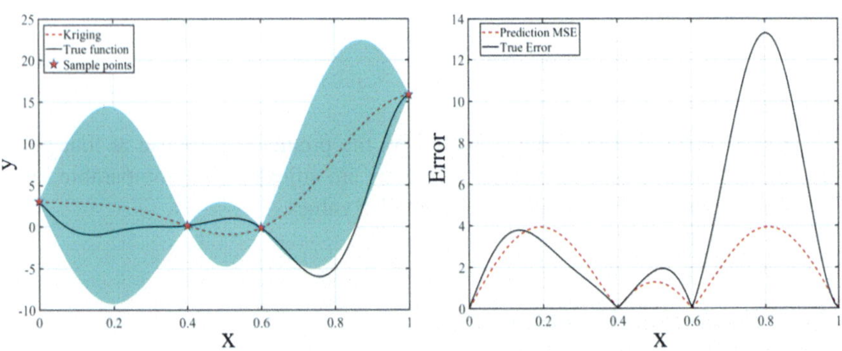

(a) Kriging approximation for a one dimensional problem

(b) Kriging prediction MSE and the true error

Fig. 5.1 Kriging prediction MSE for a one-dimensional problem

e_{mse} indicates that the surrogate is more accurate. For a large number of test points n, this error metric is equivalent to the Monte Carlo integrated MSE (IMSE) (Goel and Stander 2009; Romero et al. 2015).

The maximum MSE at all points can be used to measure the risk of a large error in prediction, and it can be formulated as follows:

$$MMSE = \max_{x \in D}[\widehat{mse}(x)] \tag{5.13}$$

where D is the design space and $\widehat{mse}(x)$ is the prediction MSE at any point within the input space. However, for problems in more than two dimensions, the areas with the maximum prediction uncertainty are generally located on the boundaries of the design domain. Thus, when this metric is used for model refinement, there is a higher chance of selecting sample points at the edges of the design domain, which may not be efficient for kriging.

5.1.5 Predictive Estimation of Model Fidelity (PEMF) Error

The PEMF method was proposed by Mehmani et al. (2015). PEMF is also a resampling-based method, and it estimates the model fidelity within the domain of interest. The PEMF method assesses the accuracy of a surrogate model by analysing the model error variation based on various numbers of training points. In PEMF, intermediate surrogates are iteratively constructed using different subsets of sample points. In each iteration, the remaining points in the whole sample set that are not used for generating the intermediate surrogate are used to estimate the median and maximum errors and determine their distributions. After all intermediate surrogate models have been constructed, a curve that reflects how the model error changes with the sample density is obtained. Since the final surrogate model is built based on the whole sample set, regression models are utilized to calculate the statistical modes of the median and maximum error distributions. If a monotonic trend (MT) criterion is satisfied by the fitted regression function for the variation of error with sample density (VESD), this VESD function is used to predict the fidelity of the final model. Otherwise, a stable implementation of k-fold CV (called PEMF-based k-fold CV) is used to predict the error of the final surrogate model. The PEMF method can be implemented using the PEMF CV toolbox (Mehmani et al. 2015).

5.2 Metrics Relying on Sampling Methods

For metrics relying on sampling methods, additional sample points are generated in the design space to enable a reasonable judgement about the surrogate model. The performance of this kind of method is strongly affected by two factors. The first factor is the number of sample points. If the number of sample points is not sufficiently large, the error estimation results may vary with different data sets, and the reliability of the model will also change. Therefore, a large number of verification samples are preferred for model verification. On the other hand, a large data set means additional simulations, which may be prohibitive when HF simulations are expensive. Thus, a reasonable number of verification samples must be chosen before testing the accuracy of a surrogate model. The second factor is the method used to generate the verification points. Generally, the random sampling method is commonly used to measure the generalization performance of a model. Latin hypercube sampling (LHS) has also been used by many researchers; the advantage of this method is that the verification samples are uniformly distributed in the design domain, enabling better estimation of the global accuracy of the surrogate model. After the samples are obtained, various types of errors between the predictions and the corresponding observations can be calculated to measure the global or local accuracy of the surrogate model. Several popular alternatives are briefly introduced in this section; for additional forms of error metrics, readers can refer to Li and Heap (2011), Franses (2016).

5.2.1 Coefficient of Determination

The coefficient of determination (R^2) is a commonly used regression-based metric for the agreement between the observed data and the predicted responses. It is usually presented as a quantity that estimates the proportion of explained variance present in the data (Grafton 2012). This metric can be calculated using the following formula (Barrett 1974; Nagelkerke 1991):

$$R^2 = 1 - \frac{\sum_{i=1}^{N_{test}} (y_i - \hat{y}_i)^2}{\sum_{i=1}^{N_{test}} (y_i - \bar{y}_i)^2} \tag{5.14}$$

where y_i is the true response, $\hat{y}_i$ is the prediction for y_i obtained with the constructed surrogate model, $\bar{y}_i$ is the mean of the true responses and N_{test} is the total number of verification samples. The same notation is also applied below for the other formulas introduced in this section. The quantity $y_i - \bar{y}_i$ can be regarded as the deviation in the worst case, i.e. the case in which there is no relation between the predictions and the observations. The expression for this metric shows that there is no assumption that the responses need to satisfy. When this metric is used as an indicator of the goodness-of-fit of a surrogate model, its values are distributed in the interval [0, 1]. The larger the value is, the closer the predicted and observed responses are. Generally, a low value ($R^2 < 0.5$) indicates that the correlation between the predicted and true values is weak, while a moderately low value ($0.5 < R^2 < 0.8$) indicates that the surrogate model may be not adequate or that there is substantial error variation (possibly caused by a large measurement spread). The main disadvantage of this method is that a higher value of R^2 may not indicate a more accurate model since the R^2 value may also increase even when irrelevant terms are added into the model. This problem can be partially alleviated by adopting a modified version of the metric (Renaud and Victoria-Feser 2010):

$$R_{adj}^2 = 1 - \frac{n-1}{n-k-1} R^2 \tag{5.15}$$

where k is the number of predictor variables in the linear regression model (if there is no intercept term, the denominator in the formula should be changed to $n - k$).

5.2.2 *Mean Square Error (MSE)*

The MSE is the mean of the overall squared prediction errors and can be expressed as

$$MSE = \frac{1}{N_{test}} \sum_{i=1}^{N_{test}} (y_i - \hat{y}_i)^2 \tag{5.16}$$

An MSE equal to zero means that the estimator can perfectly predict the response of a parameter. The smaller the MSE value is, the better the quality of the surrogate.

5.2.3 *Root Mean Square Error (RMSE)*

The RMSE can be regarded as the average vertical distance of the actual observations from the fit line (Li 2010). In this method, points with larger errors tend to be assigned higher weights by penalizing the variance. The formula for the RMSE when evaluated at a set of test points (N_{test}) is as follows:

$$RMSE = \sqrt{\frac{1}{N_{test}} \sum_{i=1}^{N_{test}} (y_i - \hat{y}_i)^2} \tag{5.17}$$

Mathematically, the RMSE is the square root of the MSE. Thus, it is also a global error metric, and a smaller value represents a higher accuracy.

5.2.4 Maximum Absolute Error (MaxAE)

The maximum absolute error (MaxAE) is a local error metric that measures the maximum approximation error of the surrogate model. The formula for the MaxAE for a set of test points can be expressed as follows (Bhattacharyya 2018):

$$MaxAE = \max(|y_i - \hat{y}_i|) \quad (i = 1, 2, \ldots, N_{test}) \tag{5.18}$$

5.2.5 Relative Maximum Absolute Error (RMAE)

The relative maximum absolute error (RMAE) is similar to the MaxAE except that it is normalized with respect to the standard deviation. The RMAE is also a metric that indicates the maximum local error of the surrogate model in the design space, although sometimes the global errors represented by other metrics are good enough. A smaller value of this metric indicates a higher local accuracy of the surrogate model. The RMAE has the following form:

$$RMAE = \frac{\max(|y_i - \hat{y}_i|)}{STD} \quad (i = 1, 2, \ldots, N_{test}) \tag{5.19}$$

5.2.6 Mean Absolute Error (MeanAE)

The mean absolute error (MeanAE) is the mean value of the errors at all verification points, with the same weight assigned to all errors. It is often regarded as the true error of the surrogate model when sufficient verification points are considered, and it is defined as follows:

$$MeanAE = \frac{1}{N_{test}} \sum_{i=1}^{N_{test}} |y_i - \hat{y}_i| \quad (i = 1, 2, \ldots, N_{test}) \tag{5.20}$$

5.2.7 Relative Average Absolute Error (RAAE)

The relative average absolute error (RAAE) was used in Jin et al. (2001) to measure the global error of a surrogate model. The smaller the value of the RAAE is, the more accurate the surrogate model. The RAAE is defined as follows:

$$RAAE = \frac{1}{N_{test} * STD} \sum_{i=1}^{N_{test}} |y_i - \hat{y}_i| \tag{5.21}$$

where STD denotes the standard deviation of the predicted responses.

5.3 Selection of Error Metrics

5.3.1 Review of Error Metrics for Surrogate Models

When multiple error metrics are available in engineering design, the question of how to select the most appropriate one arises. The performances of various error metrics have been tested in accordance with their application by several researchers. For the first of the four cases regarding the application of error metrics discussed at the beginning of this chapter, Liu et al. (2017) developed an adaptive sampling method based on the PV, representing the local error, and the CV error, representing the global error. A dynamic balance factor was used to balance local exploitation with global exploration. The proposed method was tested on several benchmark examples, and the results showed that it is more efficient compared with other alternatives and can achieve a higher accuracy with no increase in cost. Eason and Cremaschi (2014) developed a sequential sampling method that combines adaptive and space-filling characteristics. In this method, the jackknife method is used to create subsets, construct surrogate models and estimate the variance of the responses. A point with a larger PV has a higher chance to be selected as a sequential sample. When exact simulations are expensive, it may be unaffordable to calculate the true MAE and RMSE. Boopathy and Rumpfkeil (2014) proposed the maximum absolute discrepancy and the root mean square discrepancy as global and local measures, respectively, of surrogate model accuracy. Through application to a series of analytical test problems, it was found that these two metrics show good agreement with the actual MAE and RMSE. The proposed metrics were applied in a training sample selection framework, and their performances were compared by means of LHS.

For case (2), i.e. the estimation of the true fidelity of a surrogate model, Hu et al. (2018) compared the performances of LOO, BE, PEMF and PV metrics for quantifying the uncertainty of multi-fidelity models and found that in low-dimensional problems, the prediction MSE shows the best performance, while the LOO error behaves best in high-dimensional situations. Jin et al. (2001) used the R^2, RAAE and RMAE metrics to evaluate the performances of four surrogate models and found that the RBF model performed the best with a small and scarce sample set. Willmott and Matsuura (2005) compared the performances of the RMSE and MeanAE and concluded that the MeanAE is a more natural error metric than the RMSE is for reflecting the average error of a model. However, Chai and Draxler (2014) argued that the RMSE is not ambiguous in its meaning and thus is more suitable than the MAE for the case in which the model errors follow a normal distribution. Zhou et al. (2017) proposed an adaptive modelling method based on the PV obtained from the low-fidelity (LF) model and the discrepancy model, which reflects the difference between the HF and LF models. The mean of the PV over a set of points is used to represent the accuracy of the constructed surrogate model and to judge whether the stopping criterion is met.

For case (3) of the application of error metrics, i.e. selecting the optimal weights for an ensemble of surrogate models, Viana et al. (2009) discussed the performance of $PRESS_{RMS}$, which is a metric that combines the PRESS and RMSE, for model selection and found that $PRESS_{RMS}$ is a good filter for screening inaccurate surrogate models when there are sufficient samples. Goel et al. (2007) used the generalized mean square cross-validation error (GMSE) as a global metric for calculating the weights of PRS, kriging and RBF models in an ensemble of surrogate models, where the errors for each surrogate model at each sample point were considered in the heuristic formulation. Acar and Rais-Rohani (2009) also suggested using the GMSE as a global metric but proposed that the weights of the different surrogate models should be selected to minimize the overall GMSE of the ensemble, which does not depend on the pointwise accuracy. Instead of using global error measures, Sanchez et al. (2008) proposed using the PV as a local accuracy metric, with flexible weights for each component in the ensemble over the design domain. Furthermore, Acar (2010) proposed the use of pointwise CV as an alternative to the PV for local error measurement. The proposed method was tested on several problems of varying complexity, and it was found that the two methods showed similar performance.

When multiple alternative surrogate models are available and the designers want to choose the best among them, i.e. case (4), error metrics are often used to obtain an overall assessment of modelling quality. Goel and Stander (2009) applied the RMSE, PRESS and PRESS-ratio to select the best RBF network topologies and used the correlation coefficient, RMSE, ARE and MaxAE to evaluate the accuracy of the constructed surrogate model. The results showed that using the PRESS-based error metric to determine the best RBF network topology yielded the most robust result, while the PRESS-ratio-based selection criterion behaved reasonably well, but the method was very sensitive to the design of experiments (DoE) method adopted. Mao et al. (2014) proposed a new LOO validation method for selecting the best multidimensional support vector regression (SVR) model to improve the prediction performance for multi-input multi-output problems. Compared with the traditional LOO method, the proposed method is more efficient and robust. The performance of the proposed method in terms of generalization performance, numerical stability and computational cost was demonstrated on two problems. Although the LOO approach is a very popular method for model selection (Shao 1993; Yang 2007; Arlot and Celisse 2010), it is not always the first choice for all model selection problems. Gronau and Wagenmakers (2018) discussed some limitations of Bayesian LOO-CV in the context of model selection. In addition, the AIC and the Bayesian information criterion (BIC) were used to select among six stock–recruitment models, and it was found that both methods worked well in this example. When nested models are considered, however, the BIC performs better than the AIC does (Wang and Liu 2006).

5.3.2 *Performance Comparison of Commonly Used Error Metrics*

In this section, the error estimation performances of the LOO and BE metrics, which rely on testing methods, and the RMSE and MeanAE, which rely on sampling methods, are illustrated by means of two numerical examples and a cantilever beam design problem. In this test, the MeanAE results are regarded as representing the true fidelity of the surrogate model, and the other three metrics are compared with it. Several factors that may influence the evaluation results are considered, such as the number of samples, the sampling method and the noise level. Several general conclusions can be obtained from the results of the comparison, which can serve as guidance for error metric selection for problems with different characteristics. In our study, to account for the possibility of repeated points in the bootstrap method, every design variable of the inputs was normalized. Then, a small noise level between 0 and 1E-7 was added to the normalized inputs, while the responses at these points remained unchanged.

Two benchmark numerical examples, namely, the two-dimensional camelback function and the nine-variable extended Rosenbrock function (Acar 2010; Ye et al. 2018), are considered as representative low- and high-dimensional problems, respectively. The formulas for these two problems are presented below.

Function 1 (F1): Camelback (CB) function

$$\begin{gathered} f(x_1, x_2) = \left(4 - 2.1x_1^2 + \frac{x_1^4}{3}\right)x_1^2 + x_1x_2 + \left(-4 + 4x_2^2\right)x_2^2 \\ x_1 \in [-3, 3], x_2 \in [-2, 2] \end{gathered} \tag{5.22}$$

Function (F2): Extended Rosenbrock function (with nine variables)

$$f(x_1, x_2, \ldots, x_9) = \sum_{i=1}^{8} [(1 - x_i)^2 + 100(x_{i+1} - x_i^2)^2], x_i \in [-5, 10] \tag{5.23}$$

5.3.2.1 Influence of Sample Size

The influence of the sample size on the performances of the four error metrics is investigated first. The errors predicted by the different error metrics and the true errors for the problems of different dimensionalities are plotted in Fig. 5.2a–d. The number of sample points (N_{HF}) was set to 5, 10, 15 and 20 times the number of design variables, denoted by 5*ndv, 10*ndv, 15*ndv and 20*ndv, respectively, in the figures. It is obvious that as the number of sample points increases, the model

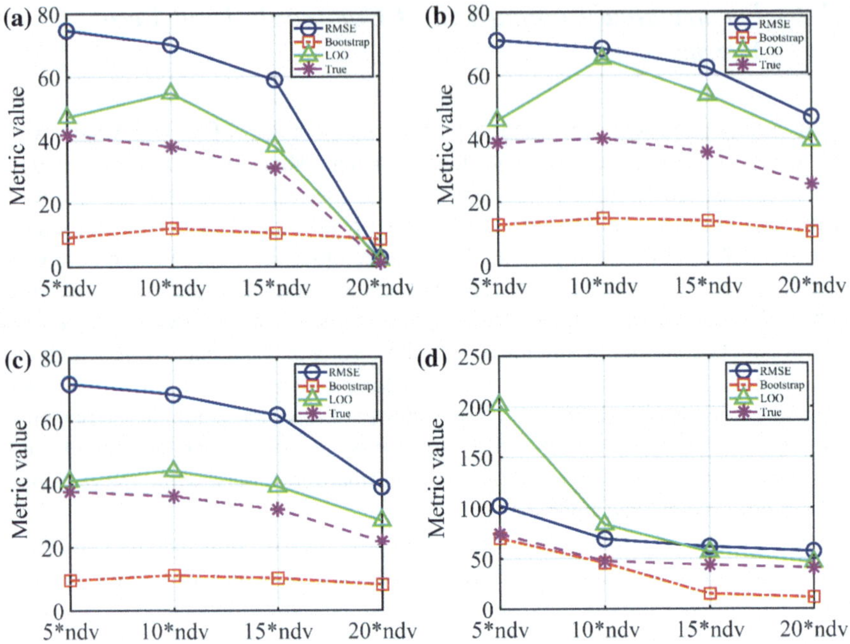

Fig. 5.2 Comparison of four error metrics with different numbers of sample points (F1): **a** kriging, **b** RBF, **c** SVR and **d** PRS

quality tends to improve. However, once the surrogate has already achieved a sufficiently high accuracy, continuing to add more sample points will contribute little to the accuracy, and the true error will remain almost the same, as shown in Fig. 5.2d. The RMSE and LOO results show a tendency similar to that of the true error, while the variation in the BE is very small for both problems even when the number of HF samples is increased to four times the initial sample size. The LOO error is always the closest to the true error in the kriging, RBF and SVR models, but it does not perform well for the PRS model when the samples are sparse, as seen in Figs. 5.2d and 5.3d. Thus, the LOO error metric is the best option for quantifying the uncertainty of kriging, RBF and SVR models. For the PRS model, the RMSE shows almost the same trend as the true error does and is very close to the true error for both F1 and F2. Thus, the RMSE is the best option for PRS models with sample sets of different sizes. It is worth noting that the RMSE overestimates the error of the PRS model as well as the LOO estimates for the other three surrogate models.

5.3.2.2 Influence of Noise Level

Due to the presence of uncertainty in engineering applications, some sampling noise may be included in the responses, which may affect the accuracy of the surrogate

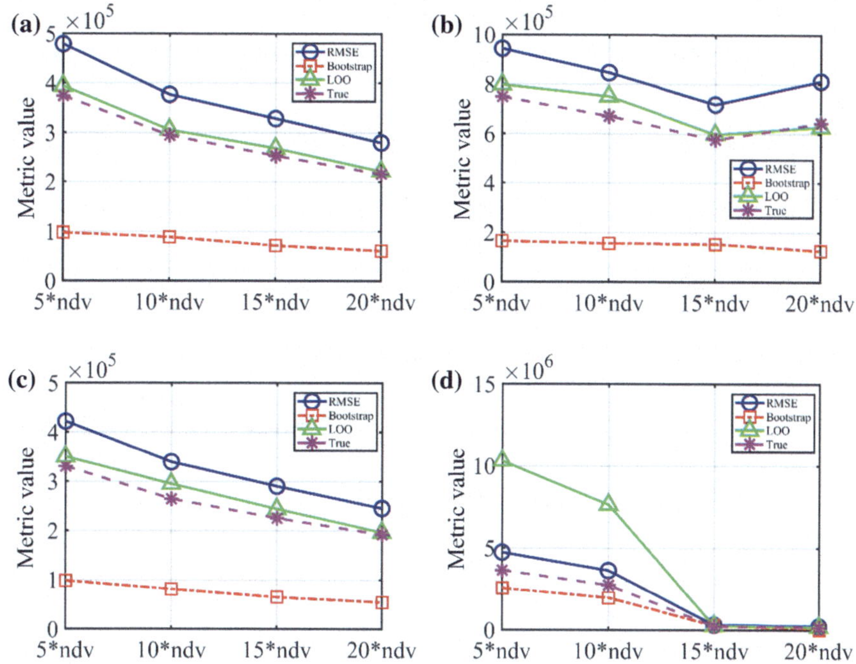

Fig. 5.3 Comparison of four error metrics with different numbers of sample points (F2): **a** kriging, **b** RBF, **c** SVR and **d** PRS

model. In this analysis, artificial noise was added to the output response value in accordance with the following formula (Zhao and Xue 2010; Zhou et al. 2016):

$$Y = f(x) + l'\delta \tag{5.24}$$

where $l' = 0 \sim 15\%$ is a scaling parameter and δ is a random number sampled from the standard Gaussian distribution $N \sim (0, 1)$. In engineering design, multiple tests with the same input parameter values need to be implemented to determine the noise level. Four levels of artificial noise, 0, 5, 10 and 15%, were added to the response values at the corresponding sample points using Eq. (5.24). The accuracy results under different levels of noise are plotted in Figs. 5.4 and 5.5.

It is obvious that in the low-dimensional case, the MeanAE values of the kriging, RBF and SVR models are very close to each other, meaning that the accuracies of these three models are similar, and that the PRS model is inferior to the other three models. Among the four error metrics, the RMSE and LOO metrics always overestimate the models, while the BE tends to underestimate the true fidelity. For the high-dimensional problem, it can be seen that the accuracies of the four surrogate models are different, but the conclusions regarding the performances of the error metrics still hold. The LOO metric almost always performs the best for the kriging, RBF and SVR models for both functions, with the exception of the results presented

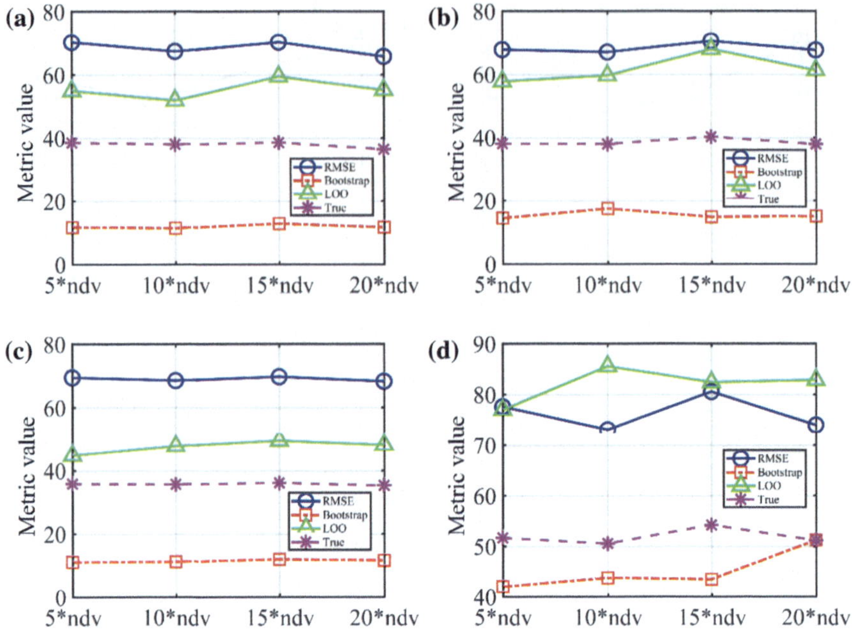

Fig. 5.4 Comparison of four error metrics under different noise levels (F1): **a** kriging, **b** RBF, **c** SVR, **d** PRS

in Fig. 5.4b, where the LOO and BE metrics show similar performance. By contrast, the bootstrap method is the most suitable verification method for the PRS model in both the high- and low-dimensional problems since the BE results show the smallest offset from the MeanAE curve. In addition, the RMSE is slightly better than the BE for all surrogate models except the PRS model, and the RMSE is superior to the LOO error for the PRS model, as seen from Figs. 5.4d and 5.5d.

5.3.2.3 Sampling Methods

In this section, two one-shot sampling methods, the LHS method and the centroidal Voronoi tessellation (CVT) method (Du et al. 1999), considered. LHS mainly exhibits the projective property, while CVT primarily satisfies the space-filling property. The number of samples was fixed to 5d for this analysis, and the calculated error metrics are presented in Figs. 5.6 and 5.7. For the problem with two design variables, it can be seen that the RMSE and MeanAE results obtained under the two different sampling methods are almost the same; thus, these two error metrics are not very sensitive to different kinds of sampling methods. The RMSE is always larger than the true error of the surrogate model. On the other hand, the results for the two testing-method-based metrics show stark differences under LHS

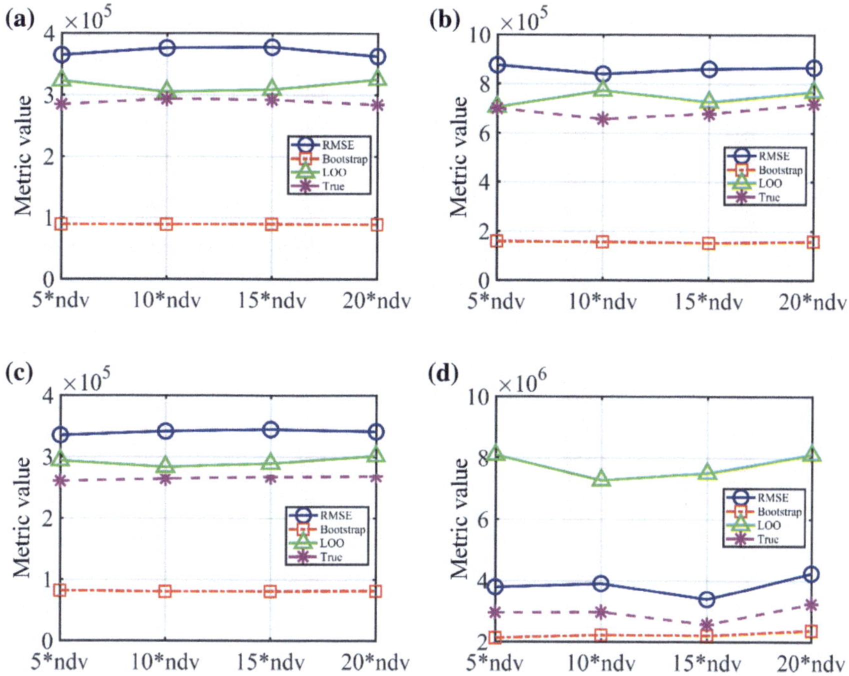

Fig. 5.5 Comparison of four error metrics under different noise levels (F2): **a** kriging, **b** RBF, **c** SVR, **d** PRS

and CVT. The LOO results obtained with LHS tend to overestimate the true error and underestimate the error obtained with the CVT sampling method, while the BE always underestimates the fidelity of the surrogate model. Thus, the performances of the LOO and bootstrap methods are easily influenced by the distribution of the sample points. For the high-dimensional problem, the conclusions regarding the LOO and BE metrics are similar to those obtained before; the only difference is that the evaluation results for all four metrics are obviously influenced by the sampling method. Therefore, it can be concluded that the LOO error usually overestimates the model error in the case of sampling methods that mainly exhibit the projective property and underestimates the accuracy of the surrogate model in the case of sampling methods that primarily satisfy the space-filling property. The RMSE is always larger than the true error, and the BE is smaller than the MeanAE, regardless of the sampling method.

5.3.2.4 Efficiency and Robustness

The time needed to evaluate the model error is another important evaluation criterion for error metrics. In this analysis, all tests were conducted on a computational

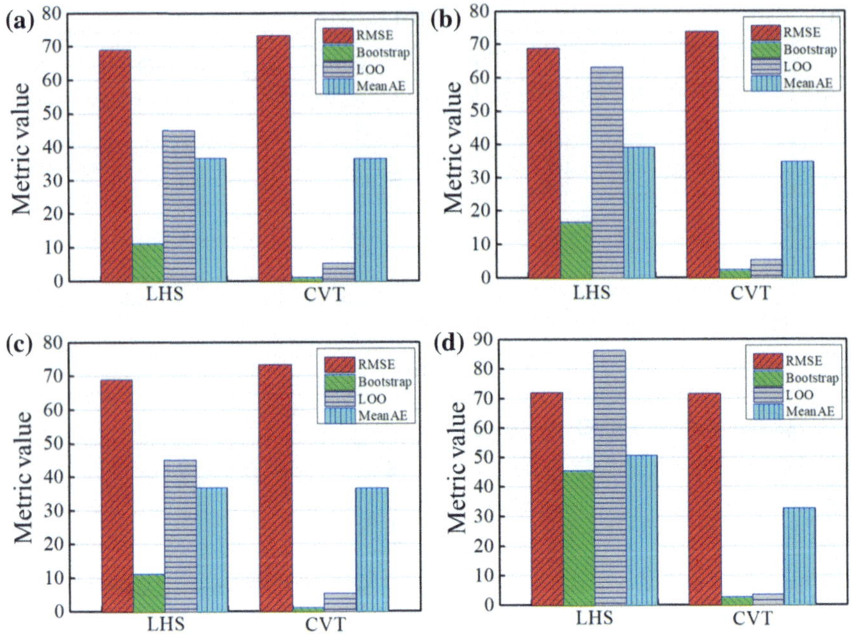

Fig. 5.6 Comparison of four error metrics under different sampling methods (F1): **a** kriging, **b** RBF, **c** SVR, **d** PRS

platform with a 3.00 GHz Intel (R) Xeon (R) E3-1220 v5 CPU and 32 GB of RAM. The number of sample points was fixed at 5d, and only the kriging approach was used to build the surrogate models here. First, 100 different sample sets were generated. Then, 100 kriging models were built based on the sample sets. Finally, the different metrics were applied to validate the accuracy of these models. The total time required to assess the accuracy of the kriging models for each method was recorded. Since error metrics relying on sampling methods require additional verification samples, 100d further sample points were randomly generated for the calculation of the RMSE and MeanAE. Most of the computational cost for the RMSE and MeanAE was incurred for generating the verification samples, and the time needed to calculate the metric value was very short or negligible. Therefore, the computational costs of the RMSE and MeanAE depend on the simulation time, and only the run times for the LOO and BE calculations are listed in Table 5.1.

The run time of the bootstrap method is approximately 36.5 times that of the LOO method for the two-dimensional problem, whereas for the high-dimensional problem, this factor changes to 11. Thus, the bootstrap method obviously incurs a higher computational cost than the LOO method does. Furthermore, the run times of the bootstrap and LOO methods increase by factors of 16.37 and 54.08,

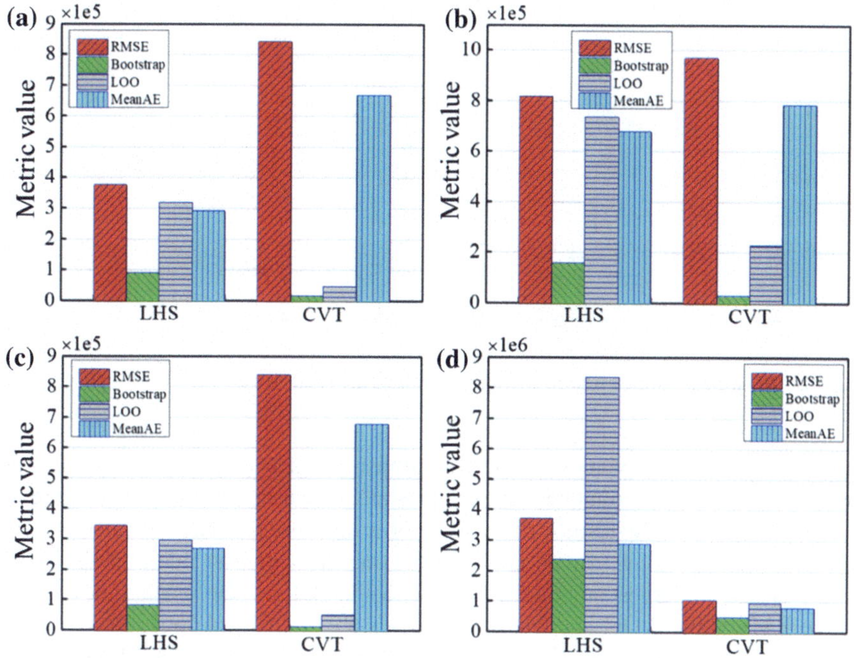

Fig. 5.7 Comparison of four error metrics under different sampling methods (F2): **a** kriging, **b** RBF, **c** SVR, **d** PRS

Table 5.1 Run times of different error metrics

	Bootstrap	LOO
F1	86.9963	2.3789
F2	1.4243e+03	1.2864e+02

respectively, when the number of design variables increases from two to nine; thus, the computational costs of these testing-method-based error metrics grow exponentially as the dimensionality of the problem increases. However, when the simulation cost is high, the time needed to build the intermediate surrogate models in the bootstrap and LOO methods is minor compared with the time required for the other methods; thus, these two methods are sometimes preferred by designers in engineering practice.

LHS may generate different sample distributions when it is run multiple times, which may result in a series of surrogate models with different accuracies. Therefore, the robustness of the results obtained with the different error metrics is also investigated here. The standard deviation of the evaluation results is selected as the index for evaluating the variation of the errors for different sample sets. The

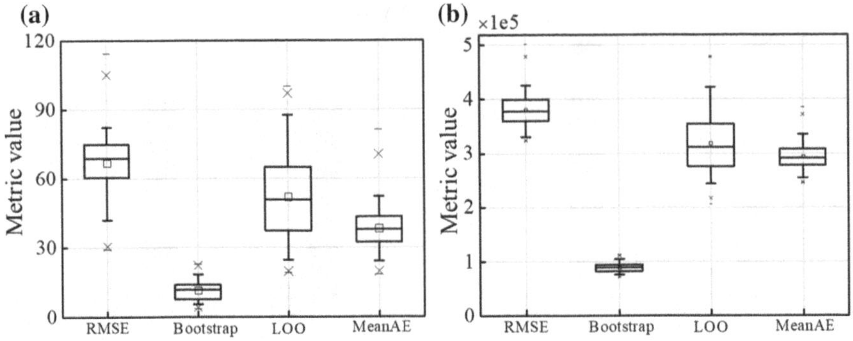

Fig. 5.8 Robustness comparison of four error metrics: **a** F1 and **b** F2

higher the standard deviation is, the less robust the metric. Box plots of the four error metrics for the two numerical examples are shown in Fig. 5.8. With regard to the median value of the results, the LOO error is the closest to the true value in both the low- and high-dimensional cases. However, the variation range of the LOO results is also the largest among the four metrics in all tests (the standard deviations of the RMSE, BE, LOO and MeanAE metrics are 13.2657, 4.0667, 19.5279 and 9.3217, respectively, for F1 and 31865.8836, 8711.2998, 54071.5815 and 25336.9871, respectively, for F2). Thus, the LOO method is the least robust under different sample distributions, while the bootstrap method is the most robust. In addition, the RMSE has the most similar standard deviation to that of the true responses, meaning that the RMSE can best reflect the variation range of the true errors.

Several conclusions can be obtained from the comparisons presented above, which can serve as guidelines for selecting error metrics for surrogate models with different characteristics, as summarized below.

Considering accuracy, efficiency and robustness, although the results of the LOO method show a larger variance, this method achieves a higher accuracy across different numbers of samples and sampling methods, is relatively fast, does not require an additional set of verification samples and does not incur any obvious drawback with an increasing sample size; therefore, it is still the best overall choice among the four methods. It should be noted that the LOO method will tend to overestimate the true fidelity of the surrogate model unless the samples used to construct the model are generated based on sample points with the space-filling property.

The RMSE is always higher than the true error, while the BE tends to underestimate the true error. When the model responses include noise, the RMSE and LOO metrics are superior to the BE for kriging, RBF and SVR models, while the BE is the best choice for PRS models. However, the bootstrap method incurs a very high computational cost compared with the LOO method and therefore may not be applicable for very high-dimensional problems or a large number of samples.

References

Acar E (2010) Various approaches for constructing an ensemble of metamodels using local measures. Struct Multidiscip Optim 42:879–896

Acar E (2015) Effect of error metrics on optimum weight factor selection for ensemble of metamodels. Expert Syst Appl 42:2703–2709

Acar E, Rais-Rohani M (2009) Ensemble of metamodels with optimized weight factors. Struct Multidiscip Optim 37:279–294

Arlot S, Celisse A (2010) A survey of cross-validation procedures for model selection. Stat Surv 4:40–79

Barrett JP (1974) The coefficient of determination—some limitations. Am Stat 28:19–20

Bhattacharyya B (2018) A critical appraisal of design of experiments for uncertainty quantification. Arch Comput Methods Eng 25:727–751

Boopathy K, Rumpfkeil MP (2014) Unified framework for training point selection and error estimation for surrogate models. AIAA J 53:215–234

Borra S, Di Ciaccio A (2010) Measuring the prediction error. A comparison of cross-validation, bootstrap and covariance penalty methods. Comput Stat Data Anal 54:2976–2989

Breiman L (1996) Heuristics of instability and stabilization in model selection. Ann Stat 24:2350–2383

Burman P (1989) A comparative study of ordinary cross-validation, v-fold cross-validation and the repeated learning-testing methods. Biometrika 76:503–514

Chai T, Draxler RR (2014) Root mean square error (RMSE) or mean absolute error (MAE)?–Arguments against avoiding RMSE in the literature. Geosci Model Dev 7:1247–1250

Du Q, Faber V, Gunzburger M (1999) Centroidal Voronoi tessellations: applications and algorithms. SIAM Rev 41:637–676

Eason J, Cremaschi S (2014) Adaptive sequential sampling for surrogate model generation with artificial neural networks. Comput Chem Eng 68:220–232

Efron B (1979) Bootstrap methods: another look at the jackknife. Ann Stat 1–26

Efron B (1983) Estimating the error rate of a prediction rule: improvement on cross-validation. J Am Stat Assoc 78:316–331

Efron B, Tibshirani R (1997) Improvements on cross-validation: the 632 + bootstrap method. J Am Stat Assoc 92:548–560

Efron B, Tibshirani RJ (1993a) An introduction to the bootstrap. Number 57 in monographs on statistics and applied probability. Chapman & Hall, New York

Efron B, Tibshirani RJ (1993b) An Introduction to the Bootstrap: monographs on Statistics and Applied Probability, vol 57. Chapman and Hall/CRC, New York and London

Franses PH (2016) A note on the mean absolute scaled error. Int J Forecast 32:20–22

Fushiki T (2011) Estimation of prediction error by using K-fold cross-validation. Stat Comput 21:137–146

Goel T, Hafkta RT, Shyy W (2009) Comparing error estimation measures for polynomial and kriging approximation of noise-free functions. Struct Multidiscip Optim 38:429

Goel T, Haftka RT, Shyy W, Queipo NV (2007) Ensemble of surrogates. Struct Multidiscip Optim 33:199–216

Goel T, Stander N (2009) Comparing three error criteria for selecting radial basis function network topology. Comput Methods Appl Mech Eng 198:2137–2150

Grafton RQ (2012) Coefficient of determination. A dictionary of climate change and the environment. Edward Elgar Publishing Limited

Gronau QF, Wagenmakers E-J (2018) Limitations of Bayesian leave-one-out cross-validation for model selection. Comput Brain Behav 1–11

Hu J, Yang Y, Zhou Q, Jiang P, Shao X, Shu L, Zhang Y (2018) Comparative studies of error metrics in variable fidelity model uncertainty quantification. J Eng Des 29:512–538

Hyndman RJ, Koehler AB (2006) Another look at measures of forecast accuracy. Int J Forecast 22:679–688

Jin R, Chen W, Simpson TW (2001) Comparative studies of metamodeling techniques under multiple modelling criteria. Struct Multidiscip Optim 23:1–13
Kohavi R (1995) A study of cross-validation and bootstrap for accuracy estimation and model selection. Montreal, Canada, Ijcai, pp 1137–1145
Larson SC (1931) The shrinkage of the coefficient of multiple correlation. J Educ Psychol 22:45
Li Y (2010) Root mean square error. In: Salkind NJ (ed) Encyclopedia of research design. Sage Publications Inc., Thousand Oaks, CA, pp 1288–1289
Li J, Heap AD (2011) A review of comparative studies of spatial interpolation methods in environmental sciences: performance and impact factors. Ecol Inform 6:228–241
Liu H, Cai J, Ong Y-S (2017) An adaptive sampling approach for kriging metamodeling by maximizing expected prediction error. Comput Chem Eng 106:171–182
Liu J, Han Z, Song W (2012) Comparison of infill sampling criteria in kriging-based aerodynamic optimization. In: 28th congress of the international council of the aeronautical sciences, pp 23–28
Mao W, Xu J, Wang C, Dong L (2014) A fast and robust model selection algorithm for multi-input multi-output support vector machine. Neurocomputing 130:10–19
Meckesheimer M, Booker AJ, Barton RR, Simpson TW (2002) Computationally inexpensive metamodel assessment strategies. AIAA J 40:2053–2060
Mehmani A, Chowdhury S, Messac A (2015) Predictive quantification of surrogate model fidelity based on modal variations with sample density. Struct Multidiscip Optim 52:353–373
Miller RG (1974) The jackknife-a review. Biometrika 61:1–15
Nagelkerke NJ (1991) A note on a general definition of the coefficient of determination. Biometrika 78:691–692
Nguyen HM, Couckuyt I, Knockaert L, Dhaene T, Gorissen D, Saeys Y (2011) An alternative approach to avoid overfitting for surrogate models. In: Proceedings of the winter simulation conference: winter simulation conference, pp 2765–2776
Picheny V (2009) Improving accuracy and compensating for uncertainty in surrogate modeling. University of Florida, Gainesville
Queipo NV, Haftka RT, Shyy W, Goel T, Vaidyanathan R, Tucker PK (2005) Surrogate-based analysis and optimization. Prog Aerosp Sci 41:1–28
Quenouille MH (1949) Approximate tests of correlation in time-series 3. In: Mathematical proceedings of the Cambridge Philosophical Society. Cambridge University Press, pp 483–484
Renaud O, Victoria-Feser M-P (2010) A robust coefficient of determination for regression. J Stat Plan Inference 140:1852–1862
Rodriguez JD, Perez A, Lozano JA (2010) Sensitivity analysis of k-fold cross validation in prediction error estimation. IEEE Trans Pattern Anal Mach Intell 32:569–575
Romero DA, Marin VE, Amon CH (2015) Error metrics and the sequential refinement of kriging metamodels. J Mech Des 137:011402
Salkind NJ (2010) Encyclopedia of research design. Sage
Sanchez E, Pintos S, Queipo NV (2008) Toward an optimal ensemble of kernel-based approximations with engineering applications. Struct Multidiscip Optim 36:247–261
Shao J (1993) Linear model selection by cross-validation. J Am Stat Assoc 88:486–494
Shao J (1996) Bootstrap model selection. J Am Stat Assoc 91:655–665
Shao J, Tu D (2012) The jackknife and bootstrap. Springer Science & Business Media
Stone M (1974) Cross-validatory choice and assessment of statistical predictions. J R Stat Soc. Series B (Methodological) 111–147
Vehtari A, Gelman A, Gabry J (2017) Practical Bayesian model evaluation using leave-one-out cross-validation and WAIC. Stat Comput 27:1413–1432
Viana FA, Haftka RT, Steffen V (2009) Multiple surrogates: how cross-validation errors can help us to obtain the best predictor. Struct Multidiscip Optim 39:439–457
Wang Y, Liu Q (2006) Comparison of Akaike information criterion (AIC) and Bayesian information criterion (BIC) in selection of stock–recruitment relationships. Fish Res 77: 220–225

Willmott CJ, Matsuura K (2005) Advantages of the mean absolute error (MAE) over the root mean square error (RMSE) in assessing average model performance. Climate Res 30:79–82

Yanagihara H, Tonda T, Matsumoto C (2006) Bias correction of cross-validation criterion based on Kullback-Leibler information under a general condition. J Multivar Anal 97:1965–1975

Yang Y (2007) Consistency of cross validation for comparing regression procedures. Ann Stat 35:2450–2473

Ye P, Pan G, Dong Z (2018) Ensemble of surrogate based global optimization methods using hierarchical design space reduction. Struct Multidiscip Optim 58:537–554

Zhao D, Xue D (2010) A comparative study of metamodeling methods considering sample quality merits. Struct Multidiscip Optim 42:923–938

Zhou Q, Shao X, Jiang P, Gao Z, Zhou H, Shu L (2016) An active learning variable-fidelity metamodelling approach based on ensemble of metamodels and objective-oriented sequential sampling. J Eng Des 27:205–231

Zhou Q, Wang Y, Choi S-K, Jiang P, Shao X, Hu J (2017) A sequential multi-fidelity metamodeling approach for data regression. Knowl-Based Syst 134:199–212

Chapter 6
Sampling Approaches

The design of experiments (DoE) is a key process in constructing a surrogate model: DoE methods are used to select the sample points at which simulations are to be conducted. Sample points generated by different DoE methods may result in the same surrogate model but with different accuracies; thus, allocating sample points reasonably in the design space is important for improving the model accuracy while respecting a certain design cost. The classic DoE methods that are widely used for physical experiments include factorial or fractional factorial methods (Box and Hunter 1961; Myers et al. 2016), central composite design (CCD) (Chen 1995; Branchu et al. 1999) and the Box–Behnken method (Box and Behnken 1960). These methods are designed to minimize the random errors caused by unknown (hidden) and/or uncontrolled variables in physical experiments. However, computer simulations are generally free of such randomness; they involve more systematic error rather than the random error encountered in physical experiments. Thus, directly applying these classic sampling methods to computational experiments may be not appropriate due to this inherent difference. In addition, Sacks et al. (1989) noted that some classic DoE methods, e.g. CCD and D-optimal design, are inefficient or even inadequate for deterministic computer simulations.

Modern DoE methods, which treat space filling of the design space as the primary consideration, have attracted widespread attention. These modern DoE methods can be generally classified into two categories: one-shot sampling methods and adaptive sampling methods. In a one-shot sampling method, the samples are all generated at the same time, and the aim of this kind of method is to fill the design space as uniformly as possible. The advantages of one-shot sampling methods are that they are simple and easy to implement. However, if the data provided by a one-shot sampling method do not meet the predefined requirements set by the designers, then a new experimental design scheme must be rearranged, and another simulation will be conducted at the new sample points, which may incur an additional computational cost. To overcome this problem, an adaptive sampling method is used to transform the one-shot sampling method into a sequential optimal

P. Jiang et al., *Surrogate Model-Based Engineering Design and Optimization*,
Springer Tracts in Mechanical Engineering,
https://doi.org/10.1007/978-981-15-0731-1_6

sampling process. The main idea is to generate a small number of samples as the initial sample set and sequentially add new sample points based on the information obtained from the existing samples. After a certain stopping criterion is reached, such as a maximum number of sample points or a predefined surrogate model accuracy, the sampling process will end, yielding the final sample set.

In this chapter, four widely used one-shot sampling methods are introduced in Sect. 6.1, and adaptive sampling methods for different kinds of surrogate models are presented in Sect. 6.2. Some novel sampling methods specifically designed for multi-fidelity surrogate models are also discussed.

6.1 One-Shot Sampling Methods

6.1.1 Uniform Design (UD)

Uniform design (UD) is a space-filling sampling method in which the selected points are uniformly distributed throughout the design domain. Suppose that there are m design variables over the domain D^m. The goal of UD is to select a set of n points $\mathbf{S}_n = \{s_1, s_2, \ldots, s_n\} \in D^m$ that are uniformly scattered in C^s. Let $M(\mathbf{S}_n)$ denote a measure of the nonuniformity of the samples; then, the goal of the sampling process is to find the sample set S^* that has the minimum $M(S)$. Let $F(S)$ denote the cumulative uniform distribution function over C^s, and let $F_n(S)$ denote the empirical cumulative distribution function of the sample set $\mathbf{S}_n$. The $\boldsymbol{L_p}$-discrepancy of the nonuniformity of $\mathbf{S}_n$ can be expressed as

$$D_p(\mathbf{S}_n) = \left[\int\limits_{C^s} |F_n(S) - F(S)|^p ds \right]^{1/p} \tag{6.1}$$

$P = 2$ is widely used; in this case, the metric becomes the $\boldsymbol{L_2}$-discrepancy. Some variants of this metric have also been developed, such as the centred $\boldsymbol{L_2}$-discrepancy (Fang and Lin 2003) and the star discrepancy (Fang et al. 2000). For the implementation of UD, readers can refer to the following website: http://www.math.hkbu.edu.hk/UniformDesign.

6.1.2 Monte Carlo Sampling (MCS)

The Monte Carlo sampling (MCS) method (also known as pseudo-random sampling) was developed by Metropolis and Ulam (Morris et al. 1995). It is still a popular sampling method for black-box functions despite its computational cost.

MCS generates pseudo-random numbers as samples and attempts to achieve space filling of the design space through its random actions (Garud et al. 2017). However, finite sets of pseudo-random numbers produced by various pseudo-random number generators may exhibit clustering or leave many regions in the domain uncovered. To solve this problem, stratified Monte Carlo sampling (SMCS) has been developed to add an element of deterministic design into the purely chaotic MCS framework. In SMCS, the design space is divided into non-random strata, and MCS is applied to each stratum. MCS and SMCS have been extensively studied and applied in various fields, e.g. mathematics (Evans and Swartz 2000), statistics (Robert and Casella 2013) and engineering design (Mordechai 2011).

6.1.3 Latin Hypercube Design (LHD)

Latin hypercube design (LHD) was first proposed by McKay et al. (2000) in 1979, and it has become one of the most well-known sampling methods for computational experiments. The LHD sampling process is described as follows. For N design variables, first construct an N-dimensional design domain $[0, 1]^N$. Then, divide each dimension into K bins of the same length $\frac{1}{K}$, thus obtaining K^N hypercubes. Finally, an N-dimensional LHD of size K can be generated from the K^N hypercubes. The K samples can be arranged in a $K \times N$ matrix $\mathbf{S} = \{\mathbf{X}_1, \mathbf{X}_2, \ldots, \mathbf{X}_K\}$, where $\mathbf{X}_i = (d_1, d_2, \ldots, d_N)\,(i = 1, 2, \ldots, K)$ is the vector of the design variables. For each column of $\mathbf{S}$, no two elements from that column may fall in the same bin. Therefore, the LHD must balance all levels of each factor in one-dimensional space. This condition on the placement of the elements in each column of L is known as the non-collapsing design condition, which reduces the original design space and allows the LHD to perform uniformly well over a range of dimensions. However, the placement across the bins is still random; consequently, the LHD approach may not ensure adequate space filling at all times. To enhance the performance of LHD, several optimization algorithms have been applied to improve its space-filling property. For an overview of these algorithms, readers can refer to Garud et al. (2017).

6.1.4 Orthogonal Array Sampling

Orthogonal array sampling (OAS) has much in common with LHD. In particular, it utilizes a process of random placement within bins similar to that of LHD. The difference is that OAS results in uniform sampling in any T-dimensional projection ($T < N$) of an N-dimensional domain, whereas LHD is restricted to $T = 1$. Thus, OAS can be regarded as a generalization of LHD. The number of sample points per bin after projection is determined by the array index λ, which has the following formula:

$$\lambda = K \times B^{-T} \tag{6.2}$$

where K is the number of sample points, B is the number of bins per dimension and T is the strength of the OAS procedure. For details of the OAS construction process, readers can refer to Owen (1992), Hedayat et al. (2012). The sampling process in OAS is determined by four predefined parameters (K, B, T and the dimensionality of the design domain, N), which make it less flexible than LHD. Moreover, the randomness in the selection of bins and the placement of samples limits the practical application of OAS. These disadvantages of OAS have been discussed by Viana et al. (2010).

6.2 Adaptive Sequential Sampling

In an adaptive sequential sampling method, new sample points are sequentially added to the original sample set without knowledge of their spatial distribution. The existing adaptive sequential sampling methods can be divided into three main types in accordance with the types of surrogate models for which they are designed, i.e. adaptive sequential sampling methods for single surrogate models, for ensembles of surrogate models and for multi-fidelity surrogate models.

6.2.1 *For Single Surrogate Models*

The adaptive sampling methods for single surrogate models are fairly simple and have enormous potential for virtually automatic implementation. Four commonly used adaptive sequential sampling methods, namely, the entropy approach, the mean square error (MSE) approach, the integrated mean square error (IMSE) approach and the cross-validation approach, are briefly introduced in this subsection.

6.2.1.1 Entropy Approach

Lindley (1956) introduced a measure of the amount of information provided by an experiment based on Shannon's entropy (Shannon 1948). The expected reduction in entropy is used as a design criterion under the Bayesian framework (Currin et al. 1988, 1991). Shewry and Wynn (1987) proved that the problem of minimizing the expected posterior entropy can be converted into the problem of maximizing the prior entropy for a discrete design space.

Suppose that T is a finite set of N sites and that design site D, associated with n samples, is evaluated to estimate the responses $y(x)$ of T, where $n<N$. After the actual responses at D are obtained, the knowledge of the predicted responses $y(x)$ depends on a normal distribution in $(N-n)$ dimensions. The predicted responses $y(x)$ are assumed to follow a Gaussian process (GP) of mean μ_D and covariance matrix C_D; the expression for such a GP is given in Sect. 6.2. In the entropy approach, an attempt is made to select a new sample set D' such that the amount of uncertainty in the predicted responses $y(x)$ is minimized.

For a GP, maximizing the amount of information obtained from the new sample set D' is equivalent to maximizing the determinant of the variance of $y(x)$. The determinant of the unconditioned covariance matrix in the Gaussian prior case has the following form:

$$\begin{aligned} \left|C_A+\tau^2 H\Sigma H^T\right| &= \begin{vmatrix} C_A+\tau^2 H\Sigma H^T & H \\ 0 & I \end{vmatrix} \\ &= \begin{vmatrix} C_A & H \\ 0 & \tau^2\Sigma H^T C_A^{-1} H + I \end{vmatrix} \\ &= |C_A|\left|\tau^2\Sigma H^T C_A^{-1} H + I\right| \\ &= |C_A|\left|H^T C_A^{-1} H + \tau^{-2}\Sigma^{-1}\right|\left|\tau^2\Sigma\right| \end{aligned} \tag{6.3}$$

where C_A is the covariance matrix of all sampling points in $A = D \cup D'$. Because $\sigma^2\Sigma$ is fixed, the problem of maximizing the above equation can be transformed into the problem of finding the design site D' that maximizes

$$|C_A|\left|H^T C_A^{-1} H + \tau^{-2}\Sigma^{-1}\right| \tag{6.4}$$

As the spatial distribution of the samples becomes increasingly diffuse, $\tau^2 \to \infty$, the problem to be solved further simplifies to the following:

$$\max\ |C_A|\left|H^T C_A^{-1} H\right| \tag{6.5}$$

However, the optimal design is highly dependent on the chosen correlation function and regression model, which are not usually determined before the analysis. One common pragmatic solution to this problem is to adopt weaker prior information than is desired in the analysis, for example, choosing a cubic model rather than a linear or quadratic polynomial model.

6.2.1.2 Mean Square Error (MSE) Approach

For a GP model, the prediction MSE at a non-test point can be calculated as

$$s^2(x)=\sigma^2[1.0 - r^T R^{-1} r + (r^T R^{-1} F - f^T)(F^T R^{-1} F)^{-1} (r^T R^{-1} F - f^T)^T] \quad (6.6)$$

The meaning of every symbol in this expression is the same as in previous expressions.

In the MSE approach, the unobserved point with the largest prediction MSE is selected as the next sequential point based on the existing GP model, i.e.

$$\max s^2(x) \quad (6.7)$$

The MSE approach has been proven to be a special case of the entropy approach in which only one new sample point is selected in each stage (Jin et al. 2002).

6.2.1.3 Integrated Mean Square Error (IMSE) Approach

The IMSE approach was proposed by Sacks et al. (1989) in 1989 for the design of computational experiments. The idea of the IMSE criterion is to determine a new sample set D' by minimizing the IMSE based on the observed samples D and the corresponding GP model, i.e.

$$\min \int s^2(x)dx \quad (6.8)$$

Note that the expression for calculating the MSE $s^2(x)$ has a similar form to that of Eq. (6.6), except that the observed data are replaced with $A = D \cup D'$ here. The hyper-parameters used to reflect the correlations between different points remain the same as those in the existing GP model.

The IMSE approach differs from the MSE approach in two respects. First, the IMSE approach uses the average MSE over the whole design space rather than the MSE at a single point. Second, in the IMSE approach, the new sample set D' is determined based on both sample sets D' and D, while in the MSE approach, it depends only on the observed sample set D.

6.2.1.4 Cross-Validation Approach

The cross-validation approach was proposed by Jin et al. (2002) to handle surrogate models that, unlike GP models, do not provide prediction errors, such as RBF models. The cross-validation approach combines the leave-one-out (LOO) method with a distance criterion. The main idea of this approach is to leave one or several samples out and reconstruct the surrogate model using only the remaining samples.

This process is repeated until all samples have been removed only once. Predicted responses at point x will be provided by both the original surrogate model ($\hat{y}(x)$) and the updated surrogate model ($\hat{y}_{-i}(x),\ (i = 1, \ldots, n)$), where $\hat{y}_{-i}(x)$ denotes the prediction of the updated surrogate model when the i-th sample in D is removed. The difference between the responses predicted by the original model and the updated surrogate model is calculated. Then, the mean square cross-validation error at all samples is used to quantify the prediction error:

$$e(x) = \sqrt{\frac{1}{n}\sum_{i=1}^{n}\left(\hat{y}_{-i}(x) - \hat{y}(x)\right)^2} \tag{6.9}$$

Furthermore, to avoid clustering of the new sample points, Eq. (6.9) can be modified to consider the distance between points:

$$\max\left[e(x) \times \min(d(x, x_i))\right] \tag{6.10}$$

where $\min(d(x, x_i))$ is the minimum distance between the next sequential sample and the existing samples.

6.2.2 *For Ensembles of Surrogate Models*

The adaptive sequential sampling process for an ensemble of surrogate models is somewhat more complex than that for a single surrogate model because it must attempt to balance the prediction uncertainties of multiple different surrogate models. Two sequential sampling criteria for an ensemble of surrogate models, namely, the identification of regions of large uncertainty and the generalized objective-oriented optimization criterion, are presented as follows.

6.2.2.1 Identification of Regions of Large Uncertainty

An important feature of a surrogate model is that it can provide prediction responses in non-tested regions. However, prediction uncertainty is also introduced during the model construction process, and different surrogate models may have different uncertainties even if they are built with the same samples. An ensemble of surrogate models can be utilized to locate regions of higher uncertainty (Goel et al. 2007). Suppose that there are N_{SM} surrogate models, the standard deviation of the prediction responses at point x is defined as

$$s_{resp}(\hat{y}(x)) = \sqrt{\frac{\sum_{i=1}^{N_{SM}}\left(\hat{y}_i(x) - \bar{y}_i(x)\right)^2}{N_{SM} - 1}} \tag{6.11}$$

where $\bar{y}_i(x)$ is the mean of the responses predicted by the N_{SM} surrogate models, that is, $\bar{y}(x) = \sum_{i=1}^{N_{SM}} \hat{y}_i(x) / N_{SM}$ and $\hat{y}_i(x)$ is the response predicted by the i-th surrogate model.

A large standard deviation of the prediction responses indicates that the surrogates are highly different in the corresponding region. Thus, additional samples should be added in this region to reduce the prediction uncertainty. It should be noted that although a high standard deviation indicates a high prediction uncertainty, a low standard deviation cannot be assumed to imply a high prediction accuracy. It may be the case that all surrogate models provide similar prediction responses, resulting in a low standard deviation of the predictions, but that all surrogate models simply behave similarly poorly in the corresponding region.

6.2.2.2 Generalized Objective-Oriented Optimization Criterion

Zhou et al. (2016) proposed an active learning multi-fidelity modelling approach based on an ensemble of surrogate models and objective-oriented sequential sampling. In this approach, kriging (Xiao et al. 2012), radial basis function (RBF) (Zhou et al. 2015a, b) and support vector regression (SVR) (Clarke et al. 2005) models are fused to map the differences between the high-fidelity (HF) and low-fidelity (LF) models. Furthermore, an active learning strategy is introduced to make full use of the already-acquired information on the difference characteristics between the HF and LF models.

Suppose there are computational codes at two levels (i.e. the HF model y_h and the LF model y_l). The HF sample set is $D_h = \{x_{h,1}, x_{h,2}, \ldots, x_{h,n_h}\}$, and the corresponding responses are $y_h = \{y_h(x_{h,1}), y_h(x_{h,2}), \ldots, y_h(x_{h,n_h})\}$. The discrepancy between the HF and LF models can be calculated as

$$c(x_{h,i}) = y_h(x_{h,i}) - y_l(x_{h,i}) \tag{6.12}$$

To make full use of the prediction capabilities of different models, the scaling function is replaced with an ensemble of surrogate models. Kriging, RBF and SVR models are selected to form the ensemble because each type of model possesses unique capabilities for handling problems with different characteristics: kriging models are useful for solving nonlinear problems, the predictions of RBF models are robust and SVR models are more accurate for problems of high dimensionality.

The ensemble is defined as the weighted average of these three surrogate models. Thus, the scaling function can be expressed as

$$\hat{c}(x) = \sum_{j=1}^{3} w_j \hat{c}_j(x) \tag{6.13}$$

where $\hat{c}(x)$ represents the predicted scaling function obtained from the ensemble and the $\hat{c}_j(x)$ $(j = 1, 2, 3)$ are the scaling functions predicted using the kriging, RBF and SVR models, respectively. The detailed forms of these three types of surrogate models have been presented in Chap. 2.

To quantify the prediction error, an error metric called the weighted cumulative error is proposed. To avoid additional simulation costs, the LOO error is used to determine the prediction accuracy based on the current sample set. The weight factors are set in accordance with the distances between each predicted point and the corresponding existing sample point, with a closer distance resulting in a larger weight. The weighted cumulative error at prediction point x_h^* is estimated as

$$\delta(x_h^*) = \sum_{i=1}^{n_h} w_i \left| \hat{c}_{-i}(x_h^*) - \hat{c}(x_h^*) \right| \tag{6.14}$$

where $\hat{c}(x_h^*)$ represents the response predicted by the ensemble of surrogate models based on the entire existing sample set $D_h = \{x_{h,1}, x_{h,2}, \ldots, x_{h,n_h}\}$, $\hat{c}_{-i}(x_h^*)$ denotes the response predicted by the ensemble of surrogate models based on the sample set $D_{h,-i} = \{x_{h,1}, \ldots, x_{i-1}, x_{i+1}, \ldots, x_{h,n_h}\}$ and w_i represents the weight factor reflecting the distance between the predicted point and the sample point $x_{h,i}$. w_i is calculated as

$$w_i = \frac{a^{-d(x_{h,i}, x_h^*)}}{a^{-\sum_{i=1}^{n_h} -d(x_{h,i}, x_h^*)}} \tag{6.15}$$

where $d(x_{h,i}, x_h^*)$ represents the Euclidean distance and a is a constant coefficient that is larger than 1.

Furthermore, to avoid sample clustering, a space-filling criterion is introduced:

$$\left\| x_h^* - x_{h,i} \right\|_2 \geq \beta \cdot \text{mean}(\min(\left\| x_{h,i} - x_{h,j} \right\|_2)), \ \forall x_{h,i}, x_{h,j} \in D_h \cap (i \neq j) \tag{6.16}$$

where β is the space-filling factor. A large value of β will cause the added sample points to be more evenly distributed, whereas a small value of β may fail to prevent the added sample points from clustering. Based on an investigation of numerous mathematical examples, the value of β is suggested to be chosen from within the interval [0.2, 0.5].

The location of the next new sample point x_h^* is determined by solving a generalized objective-oriented optimization problem, which combines the error prediction metric and the space-filling criterion. The generalized objective-oriented optimization problem is expressed as follows:

$$\begin{aligned} \max \ & \delta(x_h^*) = \sum_{i=1}^{n_h} w_i \left| \hat{c}_{-i}(x_h^*) - \hat{c}(x_h^*) \right| \\ \text{s.t.} \ & \left\| x_h^* - x_{h,i} \right\|_2 \geq \beta \cdot \text{mean}(\min(\left\| x_{h,i} - x_{h,j} \right\|_2)), \ \forall x_{h,i}, x_{h,j} \in D_h \cap (i \neq j) \end{aligned} \tag{6.17}$$

The above optimization problem can effectively improve the prediction accuracy of the model and guarantee a reasonable distance between the new sample point and the existing samples. Genetic algorithm (Coello 2000) and a penalty function method (Homaifar et al. 1994) have been widely used to find the global minimum for this optimization problem.

6.2.3 *For Multi-fidelity Models*

In this section, sequential sampling strategies for the multi-fidelity framework are introduced. For a multi-fidelity surrogate model, the selection of the sequential sample points should consider not only the location of each sample but also the level of the code to be run, which corresponds to the fidelity of the model. The selection of the next sample point depends on both the computational cost and the contribution of each code level to the prediction MSE of the final surrogate model. Three sequential adaptive sampling methods are presented in this section: the equivalent computational cost approach, which can provide the optimal combination of LF and HF samples according to the cost ratio between the HF and LF models, and the one-point-at-a-time sequential co-kriging and batch sequential co-kriging approach, which can be used to determine the location(s) and level(s) of one newly added point or several newly added points, respectively, for a co-kriging model of s levels.

6.2.3.1 Equivalent Computational Cost Approach

To solve the problems of sequentially selecting the locations of LF and HF samples for a multi-fidelity surrogate model, and determining the optimal combination of LF and HF samples given a fixed computational budget and cost ratio, Zhou et al. (2017) proposed the equivalent computational cost approach. In the proposed approach, a single HF sample or several LF samples with the same cost ratio are selected to update a multi-fidelity surrogate model. The decision is made based on which choice has the greater ability to improve the prediction accuracy.

In the equivalent computational cost approach, the multi-fidelity surrogate model is built based on an additive LF GP model and a discrepancy GP model based on the discrepancies between the LF and HF responses. The multi-fidelity surrogate model can be expressed as

$$\hat{y}_h(x) = \hat{y}_l(x) + \hat{\delta}(x) \tag{6.18}$$

where the GP models $\hat{y}_l(x)$ and $\hat{\delta}(x)$ are assumed to be independent.

The mean prediction of the multi-fidelity surrogate model can be obtained as

$$m(\hat{y}_h(x)) = E(\hat{y}_l(x)|y_l) + E(\hat{\delta}(x)|(y_h - E(\hat{y}_l(x)|y_l))) \tag{6.19}$$

and the corresponding prediction MSE has the following form:

$$s^2(\hat{y}_h(x)) = s^2(\hat{y}_l(x)|y_l) + s^2(\hat{\delta}(x)|(y_h - E(\hat{y}_l(x)|y_l))) \tag{6.20}$$

where $s^2(\hat{y}_l(x)|y_l)$ and $s^2(\hat{\delta}(x)|(y_h - E(\hat{y}_l(x)|y_l)))$ are the prediction MSEs of the LF GP model and the discrepancy GP model, respectively.

The average MSE over a set of points is used to quantify the prediction error of the multi-fidelity surrogate model, which is calculated as

$$e_0 = \frac{1}{t}\sum_{i=1}^{t}\sqrt{s^2(\hat{y}_h(x_i))} \tag{6.21}$$

where $x_i \in D_t = \{x_1, x_2, \ldots, x_t\}$ is a point in a randomly generated test set within the design space. This error metric is equivalent to the Monte Caro IMSE if there are sufficient test points.

Step 1: Estimate the predicted level of improvement with LF samples

The potential prediction improvement PIL_l with the addition of q LF samples, where q is a predefined cost ratio between the HF and LF models, is estimated by updating the multi-fidelity surrogate model. The q potential LF samples are added sequentially at the location where the prediction error is highest for the previous surrogate model. The multi-fidelity surrogate model is updated after the selection of every potential sample point. The sequential criterion can be expressed as follows:

$$\begin{aligned} &\text{find } x_l^* \\ &\max\ s^2(\hat{y}_h(x_l^*)) \\ &\text{s.t. } \left\|x_l^* - x_{l,i}\right\| \geq \text{cluster threshold}, x_{l,i} \in D_l \end{aligned} \tag{6.22}$$

The constraint is imposed to avoid clustering of the sample data. For every existing sample, the minimum Euclidean distance with respect to the other sample data is calculated. The clustering threshold is set to half of the average Euclidean distance.

During the sequential sampling process, for efficiency, the predicted responses at the selected LF sample points are treated as the actual responses when updating the correlation matrix in the multi-fidelity surrogate model. Therefore, there is no need to obtain the true responses at the LF sample points during the sequential sampling process. The sequential criterion is applied by means of a genetic algorithm.

After the q LF samples have been sequentially added to the LF sample set D_l, the updated average prediction MSE of the test set D_t is recalculated as follows:

$$e_l = \frac{1}{t}\sum_{i=1}^{t}\sqrt{s^2(\hat{y}_h(x_i))} \tag{6.23}$$

The prediction improvement level PIL_l is defined as

$$PIL_l = (e_0 - e_l)/e_0 \times 100\% \tag{6.24}$$

The detailed process of obtaining PIL_l is presented in Algorithm 6.1.

Algorithm 6.1. Search the sequential LF samples and determine the PIL_l
Input: The multi-fidelity sampling data, cost ratio q, test set D_t and current prediction error e_0
Output: The added LF samples and improvement level PIL_l
1 Begin
2 While $(i \leq q)$ do
3 $x_{l,i}^* \leftarrow$ Find $x_{l,i}^*$ to satisfy the sequential criterion
4 $D_l \leftarrow$ Update the LF input set by adding $x_{l,i}^*$ into original D_l
5 $\hat{y}_{l,i}^* \leftarrow$ Obtain the prediction response at sampling point $x_{l,i}^*$
6 $Y_l \leftarrow$ Update the LF sampling set by adding $\hat{y}_{l,i}^*$ into Y_l
7 $\hat{y}_h(x) \leftarrow$ Update the multi-fidelity surrogate model
8 $i = i + 1$
9 end while
10 $e_l \leftarrow$ Calculate the e_l for the updated multi-fidelity surrogate model
11 $PIL_l \leftarrow$ Calculate the prediction improvement level
12 end

Step 2: Estimate the predicted level of improvement with an HF sample

To locate the next HF sample point, the following optimization problem is utilized:

$$\begin{aligned} &\text{find } x_h^* \\ &\max\ s^2(\hat{y}_h(x_h^*)) \\ &\text{s.t. } \left\| x_h^* - x_{h,i} \right\| \geq \text{cluster threshold},\ x_{h,i} \in D_h \end{aligned} \tag{6.25}$$

which is similar to the problem for the selection of LF samples except that the distance constraint is replaced by the distance between the next sequential point x_h^* and the samples in the HF sample set D_h.

A genetic algorithm is again used to solve the above optimization problem. The multi-fidelity surrogate model will be updated after the location of the HF sample is determined. In contrast to the addition of q LF samples, only one HF sample will be added. Then, the average prediction MSE e_h for the test set D_t will be calculated based on the updated multi-fidelity surrogate model.

The prediction improvement level PIL_h is calculated as

$$PIL_h = (e_0 - e_h)/e_0 \times 100\% \tag{6.26}$$

Algorithm 6.2 presents a detailed illustration of the process of obtaining the location of the HF sample and calculating the prediction improvement level PIL_h.

Algorithm 6.2. Search the sequential HF samples and determine the PIL_h
Input: The multi-fidelity sampling data, test set D_t and current prediction error e_0
Output: The added HF sample and improvement level PIL_h
1 Begin
2 $x^*_{h,i}$ ← Find $x^*_{h,i}$ to satisfy the optimization criterion
3 D_h ← Update the HF input set by adding $x^*_{h,i}$ into original D_h
4 $\hat{y}^*_{h,i}$ ← Obtain the prediction response at sampling point $x^*_{h,i}$
5 Y_h ← Update the HF sampling set by adding $\hat{y}^*_{h,i}$ into Y_h
6 $\hat{y}_h(x)$ ← Update the multi-fidelity surrogate model
7 e_h ← Calculate the e_h for the updated multi-fidelity surrogate model
8 PIL_h ← Calculate the prediction improvement level
9 end

Once PIL_l and PIL_h have been obtained, they are compared to determine whether one HF sample or q LF samples should be added. If $PIL_l > PIL_h$, then the LF samples are determined to contribute more to improving the prediction accuracy of the multi-fidelity surrogate model, and the obtained q sample points will be added to the LF sample set. Otherwise, the addition of one HF sample is preferred. The equivalent computational cost approach can assist in constructing a multi-fidelity surrogate model of high accuracy by identifying the optimal HF-to-LF sample size ratio and the optimal sample locations.

6.2.3.2 One-Point-at-a-Time Sequential Co-Kriging

This criterion was proposed by Gratiet et al. (Gratiet and Cannamela 2012; Le Gratiet and Cannamela 2015) for determining the locations of sequential sample points for a co-kriging model of s levels. Code level 1 corresponds to the model that is the most inaccurate, while code level s denotes the model with the highest accuracy.

For a model of code level s, a new point x_{new} can be selected by solving the following optimization problem:

$$x_{new} = \arg\max_{x} k^s_{n_s}(x) \tag{6.27}$$

where $k_s(x)$ is the prediction variance at point x. Thus, the next simulation is conducted at the point where the prediction MSE of the model is at its maximum. For two models of successive code levels $l - 1$ and l, the sequential procedure is modified to determine at which of the two levels the next simulation should be conducted.

For a co-kriging model constructed from GP models δ_l $(l = 1, \ldots, s)$, the prediction MSE of model δ_l at point x can be written as

$$k_{n_l}^{l}(x) = \rho_{l-1}^{2} k_{n_{l-1}}^{l-1}(x) + \sigma_{\delta_l}^{2}(x) \tag{6.28}$$

and

$$\sigma_{\delta_l}^{2}(x) = \sigma_l^2 \left(1 - \begin{pmatrix} h_l(x)^T & r_l(x)^T \end{pmatrix} \begin{pmatrix} 0 & H_l^T \\ H_l & R_l \end{pmatrix}^{-1} \begin{pmatrix} h_l(x) \\ r_l(x) \end{pmatrix} \right) \tag{6.29}$$

where the meanings of the correlation matrix $h_l(x)$, $r_l(x)$, H_l and R_l are the same as in Chap. 4. It can also be easily found that $\sigma_{\delta_1}^2 = k_{n_1}^1(x, x)$. Then, the prediction MSE of $\hat{y}_l(x)$ $(l = 1, \ldots, s)$ can be expressed as

$$k_{n_l}^{l}(x) = \sum_{i=1}^{l} \sigma_{\delta_i}^{2}(x) \prod_{j=i}^{l-1} \rho_j^2 \tag{6.30}$$

where the kernels $\{r_l(x, x'; \theta_l)\}_{i=1,\ldots,s}$ are functions of the known hyper-parameters $\{\theta_l\}_{l=1,\ldots,s}$, which represent the characteristic length scales of the kernels (Rasmussen 2004) and are often calculated using the maximum likelihood estimation method. The IMSE at level l can be written as

$$IMSE^l = \int_Q k_{n_l}^l(x)\, dx \tag{6.31}$$

Then, the reduction in the IMSE when a new point x_{new} is added at level l can be approximated as follows:

$$IMSE_{red}^{l}(x_{new}) = \sum_{i=1}^{l} \sigma_{\delta_i}^{i}(x_{new}) \prod_{j=i}^{l-1} \rho_j^2 \prod_{k=1}^{m} \theta_i^k \tag{6.32}$$

where $\theta_l = (\theta_l^1, \theta_l^2, \ldots, \theta_l^m)$. The first factor, $\sigma_{\delta_i}^{i}(x) \prod_{j=i}^{l-1} \rho_j^2$, represents the contribution of the prediction error of GP model δ_i to the variance of the co-kriging model. The second factor, $\prod_{k=1}^{m} \theta_i^k$, reflects the influence of the new point x_{new} on the GP model of level l. It has been proven that the reduction in $\bar{\sigma}_{\delta_l}^2 = \int_Q \sigma_{\delta_l}^2(x) dx$ is of the same order of magnitude as the product of $\sigma_{\delta_i}^2(x_{new})$ and $\prod_{k=1}^{m} \theta_i^k$.

Suppose that the cost ratio between the models of code levels l and $l-1$ is $C_{l/l-1}$. The model of level $l-1$ is more worthwhile to run when $C_{l/l-1} IMSE_{red}^{l-1}(x_{new}) > IMSE_{red}^{l}(x_{new})$. Otherwise, the model of level l (i.e. $y_l(x)$) will be selected. Based on this criterion, the one-at-a-time sequential co-kriging process is specified as shown in Algorithm 6.3. This process can balance the effects of both the computational cost and the contributions of models at different levels to the co-kriging variance.

Algorithm 6.3 one-point-at-a-time sequential Co-kriging
Find x_{new} by maximizing $k_{n_s}^{s}(x)$
for $l = 2, \ldots, s$ do
if $(\sigma_{\delta_l}^2(x) < \bar{\sigma}_{\delta_l}^2)$ then
Run $y_{l-1}(x_{new})$ end for
else
if $(C_{l/l-1} IMSE_{red}^{l-1}(x_{new}) > IMSE_{red}^{l}(x_{new}))$ then
Run $y_{l-1}(x_{new})$ end for
end if
end if
end for
if $(l = s)$ then
Run $y_l(x_{new})$
end if

Two models of successive levels $l-1$ and l are evaluated in Algorithm 6.3. The sequential procedure begins at levels one and two and then proceeds to levels two and three, and so on. The loop stops if the model at level $l-1$ is more promising than the model at level l, and the responses $y_1(x_{new}), \ldots, y_{l-1}(x_{new})$ are evaluated. The algorithm tends to estimate models of lower fidelity because the loop starts at level 1 and proceeds to level s. There is no need to update the hyper-parameters $\{\theta_l, \sigma_l^2\}_{l=1,\ldots,s}$ during the loop; they are estimated based on maximum likelihood estimation when the loop stops. The purpose of the first test criterion, $\sigma_{\delta_l}^2(x) < \bar{\sigma}_{\delta_l}^2$, is to check whether the point x_{new} is worthwhile to evaluate for the model of level l. The second criterion, $C_{l/l-1} IMSE_{red}^{l-1}(x_{new}) > IMSE_{red}^{l}(x_{new})$, is used to check which model is more promising between code levels $l-1$ and l. If the model at level l has greater potential than the model at level $l-1$, it will be compared with the model at the following code level $l+1$. Algorithm 6.3 is repeated until the desired prediction accuracy of the co-kriging model is reached or the computational budget is exhausted.

Above, it is assumed that the computational burden monotonically increases with the model level. However, this assumption can be abandoned; in this case, the models at all levels need to be evaluated instead of stopping as soon as level $l-1$ is more promising than level l. In this situation, the most promising model can be determined by minimizing the following quantity:

$$\prod_{i=l}^{s} C_{i+1/i} IMSE_{red}^{i}(x_{new}) \tag{6.33}$$

where $C_{s+1/s} = 1$. This quantity represents the potential reduction in the prediction uncertainty at code level l given the same computational cost as that of obtaining a sample at the highest level. It should be noted that the MSE may not reflect the true error of the model; in this case, the cross-validation error can be used. For more details, readers can refer to Le Gratiet and Cannamela (2015).

6.2.3.3 Batch Sequential Co-Kriging

In this section, the sequential sampling process for a co-kriging model is extended to consider the addition of q points at a time (Gratiet and Cannamela 2012; Le Gratiet and Cannamela 2015). The principle is demonstrated as follows. q_l new samples are selected for the model of the l - th fidelity level; the details of the method used to select the samples are shown in Algorithm 6.4. Then, these points are assumed to be known for models $y_{l-1}(x), \ldots, y_1(x)$, and q_{l-1} new sample points for model y_{l-1} are determined using the same method. It should be noted that the co-kriging variances are computed independently of the observations, and there is no need to evaluate $y_{l-1}(x), \ldots, y_1(x)$ at the q_l new sample points. Next, the sequential process loops from level $l-2$ to level 1. Finally, a total of $\sum_{i=j}^{l} q_i$ points are added to level j during the loop. The locations of these points $\{q_1, q_2, \ldots, q_l\}$ are determined as those that offer the largest potential uncertainty reduction subject to a predefined computational budget limitation. The total computational burden here is

$$T = \sum_{j=1}^{l} \sum_{i=j}^{l} q_i t_j \tag{6.34}$$

where t_j $(j = 1, \ldots, s)$ represents the computational cost of evaluating $y_j(x)$ $(j = 1, \ldots, s)$. The procedure for adding q points at a time is shown in Algorithm 6.4.

Algorithm 6.4 $(q_i)_{i=1,\ldots,s}$ points at-a-time sequential co-kriging
1 Allocate $\{q_1, \ldots, q_l\}$ such that $\sum_{j=1}^{l} \sum_{i=j}^{l} q_i t_j$ is equal to predefined budget T
2 Set $(N_{MCMC}^{i})_{i=1,\ldots,t}$ for the Metropolis–Hastings (M-H) procedures
3 Generate N_{MCMC}^{l} samples with respect to $k_{n_l}^{l}(x)$
4 Solve $N^l = \max_{N \geq q_l} \min_{x \in (C_i^l)_i} s_l^2(x)$ to find the N^l cluster $(C_i^l)_{i=1,\ldots,N^l}$
5 Select from $(C_i^l)_{i=1,\ldots,N^l}$ the q_l points $(x_{new,i}^{l})_{i=1,\ldots,q_l}$ by maximizing $k_{n_l,adj}^{l}(x)$
6 for $m = l-1, \ldots, 1$ do
7 Compute $k_{n_m + \sum_{i=m+1}^{l} q^i}^{m}(x)$ after adding the new points $((x_{new,i}^{j})_{i=1,\ldots,q_j})_{j=m+1,\ldots,l}$

(continued)

(continued)

Algorithm 6.4 $(q_i)_{i=1,..,s}$ points at-a-time sequential co-kriging
8 Generate N^m_{MCMC} samples with respect to $k^m_{n_m+\sum_{i=m+1}^{l} q^i}(x)$
9 Find the N^m cluster centres $(C^m_i)_{i=1,\dots,N^m}$ such that $N^m = \max_{N\geq q_m} \min_{x\in(C^m_i)_i} k^m_{n_m+\sum_{i=m+1}^{l} q^i}(x)$
10 Select from $(C^m_i)_{i=1,\dots,N^m}$ the q_m points $(x^m_{new,i})_{i=1,\dots,q_m}$ by maximizing $k^m_{n_m+\sum_{i=m+1}^{l} q^i,adj}(x)^*$
11 end for

The term $k^m_{n_m+\sum_{i=m+1}^{l} q^i}(x)$ in Algorithm 6.4 represents the prediction MSE of $Y^m_{n_m}(x)$ conditioned on $((x^j_{new,i})_{i=1,\dots,q_j})_{j=m+1,\dots,l}$ when the hyper-parameters $(\sigma_i^2)_{i=1,\dots,l}$ and $(\theta_i)_{i=1,\dots,l}$ are treated as known. Moreover, $k^m_{n_m+\sum_{i=m+1}^{l} q^i}(x)^*$ is the adjustment to $k^m_{n_m+\sum_{i=m+1}^{l} q^i}(x)$ when leave-one-out cross-validation (LOO-CV) is adopted. The expression for $k^m_{n_m+\sum_{i=m+1}^{l} q^i}(x)^*$ has the following form:

$$k^m_{n_m+\sum_{i=m+1}^{l} q^i}(x)^* = \sum_{i=1}^{l} \sigma^2_{\delta_i}(x_{new})^* \left(1+\sum_{j=1}^{n_i} \frac{(\varepsilon_{LOO-CV,i}(x_{i,j}) - \hat{\rho}_{i-1}\varepsilon_{LOO-CV,i-1}(x_{i-1,j}))^2}{s^2_{LOO-CV,i}(x_{i,j}) - \hat{\rho}^2_{i-1}s^2_{LOO-CV,i-1}(x_{i-1,j})}\right) \times \prod_{j=i}^{l-1}\rho_j^2 \prod_{k=1}^{m}\theta_i^k \tag{6.35}$$

where $\varepsilon_{LOO-CV,i}(x_{i,j})$ and $s^2_{LOO-CV,i}(x_{i,j})$ represent the LOO-CV error and variance, respectively. A Gaussian jumping distribution with the corresponding standard deviation is used for the Metropolis–Hastings procedure, and the acceptance rate is approximately 30% (Roberts et al. 1997). The specific definitions of $\varepsilon_{LOO-CV,i}(x_{i,j})$ and $s^2_{LOO-CV,i}(x_{i,j})$ are provided as follows.

For model level l, the models at lower levels are considered to satisfy $D_l \subseteq D_{l-1} \subseteq \cdots \subseteq D_1$. ξ_j is defined as the index of D_j corresponding to the i-th point $x_{l,i}$ of D_l, with $1 \leq j \leq l$. The LOO-CV error $\varepsilon_{LOO-CV,l}(x_{i,j})$ at level l and point $x_{l,i}$ can be calculated as follows:

$$\varepsilon_{LOO-CV,l}(x_{l,i}) = \hat{\rho}_{l-1}\varepsilon_{LOO-CV,l-1}(x_{l,i}) + \left[R_l^{-1}\left(y_l - H_l\begin{pmatrix}\hat{\rho}_{l-1}\\ \hat{\beta}_l\end{pmatrix}\right)\right]_{\xi_l} / \left[R_l^{-1}\right]_{\xi_l,\xi_l} \tag{6.36}$$

where

$$\begin{pmatrix}\hat{\rho}_{l-1}\\ \hat{\beta}_l\end{pmatrix} = \left(H^T_{l,-\xi_l}K_l H_{l,-\xi_l}\right)^{-1} H^T_{l,-\xi_l}K_l y_{l,-\xi_l} \quad \text{and}$$

$$K_l = \left[R_l^{-1}\right]_{-\xi_l,-\xi_l} - \left[R_l^{-1}\right]_{-\xi_l,\xi_l}\left[R_l^{-1}\right]_{\xi_l,-\xi_l} / \left[R_l^{-1}\right]_{\xi_l,\xi_l}$$

Similarly, the LOO-CV variance $s^2_{LOO-CV,l}(x_{i,j})$ can be computed as follows:

$$s^2_{LOO-CV,l}(x_{l,i}) = \hat{\rho}^2_{l-1} s^2_{LOO-CV,l-1}(x_{l,i}) + s^2_{\delta_l,-\xi_l} / \left[R_l^{-1}\right]_{\xi_l,\xi_l} + \varsigma_l \tag{6.37}$$

with $\varsigma_l = u_l^2 \left(H^T_{l,-\xi_l} K_l H_{l,-\xi}\right)^{-1}$, $u_l = \left[R_l^{-1} H_l\right]_{\xi_s} / \left[R_l^{-1}\right]_{\xi_l,\xi_l}$ and

$$\sigma^2_{l,-\xi_l} = \frac{\left(y_{l,-\xi_l} - H_{l,-\xi_l}\begin{pmatrix}\hat{\rho}_{l-1}\\ \hat{\beta}_l\end{pmatrix}\right)^T K_s \left(y_{l,-\xi_l} - H_{l,-\xi_l}\begin{pmatrix}\hat{\rho}_{l-1}\\ \hat{\beta}_l\end{pmatrix}\right)}{n_l - p_l - 2} \tag{6.38}$$

where $H_{l,-\xi_l} = \left[y_{l,-\xi_l} \quad F_{l,-\xi_l}\right]$.

Moreover, the following quantity is considered:

$$IMSE_{red,q} = \sum_{i=1}^{l} \sum_{r=1,\ldots,q_i} \sigma^2_{\delta_i}(x^i_{new,r}) \prod_{j=i}^{l-1} \rho_j^2 \prod_{k=1}^{m} \theta_i^k \tag{6.39}$$

The allocation $\{q_1, \ldots, q_l\}$ is determined by solving the following optimization problem:

$$\{q_1, \ldots, q_l\} = \arg \max_{\{q_1,\ldots,q_l\}} IMSE_{red,q} \text{ such that } \sum\nolimits_{j=1}^{l} \sum\nolimits_{i=j}^{l} q_i t_j = T \tag{6.40}$$

This optimization problem is usually very complex to solve, and suboptimal solutions are often adopted. The possible allocations can be fully explored when the number of code levels and the budget are relatively small.

References

Box GE, Behnken DW (1960) Some new three level designs for the study of quantitative variables. 2:455–475

Box GE, Hunter JS (1961) The 2 k − p fractional factorial designs 3:311–351

Branchu S, Forbes RT, York P, Nyqvist H (1999) A central composite design to investigate the thermal stabilization of lysozyme. 16:702–708

Chen W (1995) A robust concept exploration method for configuring complex systems. Georgia Institute of Technology

Clarke SM, Griebsch JH, Simpson TW (2005) Analysis of support vector regression for approximation of complex engineering analyses. J Mech Des 127:1077–1087

Coello CAC (2000) Use of a self-adaptive penalty approach for engineering optimization problems. Comput Ind 41:113–127

Currin C, Mitchell T, Morris M, Ylvisaker D (1988) A Bayesian approach to the design and analysis of computer experiments. Oak Ridge National Lab, TN (USA)

Currin C, Mitchell T, Morris M, Ylvisaker D (1991) Bayesian prediction of deterministic functions, with applications to the design and analysis of computer experiments. J Am Stat Assoc 86:953–963

Evans M, Swartz T (2000) Approximating integrals via Monte Carlo and deterministic methods. OUP Oxford

Fang K-T, Lin DKJ (2003) Ch. 4. Uniform experimental designs and their applications in industry. 22:131–170

Fang K-T, Lin DK, Winker P, Zhang Y (2000) Uniform design: theory and application. 42:237–248

Garud SS, Karimi IA, Kraft M (2017) Design of computer experiments: a review. 106:71–95

Goel T, Haftka RT, Shyy W, Queipo NV (2007) Ensemble of surrogates. Struct Multidiscip Optim 33:199–216

Gratiet LL, Cannamela C (2012) Kriging-based sequential design strategies using fast cross-validation techniques with extensions to multi-fidelity computer codes. arXiv:1210.6187

Hedayat AS, Sloane NJA, Stufken J (2012) Orthogonal arrays: theory and applications. Springer Science & Business Media

Homaifar A, Qi CX, Lai SH (1994) Constrained optimization via genetic algorithms. Simulation 62:242–253

Jin R, Chen W, Sudjianto A (2002) On sequential sampling for global metamodeling in engineering design. In: ASME 2002 international design engineering technical conferences and computers and information in engineering conference, pp 539–548. American Society of Mechanical Engineers

Le Gratiet L, Cannamela C (2015) Cokriging-based sequential design strategies using fast cross-validation techniques for multi-fidelity computer codes. Technometrics 57:418–427

Lindley DV (1956) On a measure of the information provided by an experiment. Ann Math Stat 986–1005

McKay MD, Beckman RJ, Conover WJ (2000) A comparison of three methods for selecting values of input variables in the analysis of output from a computer code. 42:55–61

Mordechai S (2011) Applications of Monte Carlo method in science and engineering

Morris MD, Mitchell TJ (1995) Exploratory designs for computational experiments. 43:381–402

Myers RH, Montgomery DC, Anderson-Cook CM (2016) Response surface methodology: process and product optimization using designed experiments. Wiley

Owen AB (1992) Orthogonal arrays for computer experiments, integration and visualization. 439–452

Rasmussen CE (2004) Gaussian processes in machine learning. Advanced lectures on machine learning. Springer, pp 63–71

Robert C, Casella G (2013) Monte Carlo statistical methods. Springer Science & Business Media

Roberts GO, Gelman A, Gilks WR (1997) Weak convergence and optimal scaling of random walk Metropolis algorithms. Ann Appl Probab 7:110–120

Sacks J, Welch WJ, Mitchell TJ, Wynn HP (1989) Design and analysis of computer experiments. Stat Sci 409–423

Shannon CE (1948) A mathematical theory of communication. Bell Syst Tech J 27:379–423

Shewry MC, Wynn HP (1987) Maximum entropy sampling. J Appl Stat 14:165–170

Viana FA, Venter G, Balabanov V (2010) An algorithm for fast optimal Latin hypercube design of experiments. 82:135–156

Xiao M, Gao L, Shao X, Qiu H, Jiang P (2012) A generalised collaborative optimisation method and its combination with kriging metamodels for engineering design. J Eng Des 23:379–399

Zhou Q, Shao X, Jiang P, Zhou H, Cao L, Zhang L (2015a) A deterministic robust optimisation method under interval uncertainty based on the reverse model. J Eng Des 26:416–444

Zhou Q, Shao X, Jiang P, Zhou H, Shu L (2015b) An adaptive global variable fidelity metamodeling strategy using a support vector regression based scaling function. Simul Model Pract Theory 59:18–35

Zhou Q, Shao X, Jiang P, Gao Z, Zhou H, Shu L (2016) An active learning variable-fidelity metamodelling approach based on ensemble of metamodels and objective-oriented sequential sampling. J Eng Des 27:205–231

Zhou Q, Wang Y, Choi S-K, Jiang P, Shao X, Hu J (2017) A sequential multi-fidelity metamodeling approach for data regression. Knowl Based Syst 134:199–212

Chapter 7
Surrogate-Model-Based Design and Optimization

7.1 Surrogate-Model-Based Deterministic Design Optimization

Since most engineering design problems involve time-consuming simulations and analysis, surrogate models are often used for fast calculations, sensitivity analysis, exploring the design space and supporting optimal design (Sacks et al. 1989; Gutmann 2001; Hsu et al. 2003; Clarke et al. 2005; Martin and Simpson 2005; Chen and Cheng 2010; Sun et al. 2011; Kitayama et al. 2013; Long et al. 2015). Surrogate models are also widely used in design optimization (Gu 2001; Simpson et al. 2004; Wang and Shan 2007; Forrester and Keane 2009; Viana et al. 2014). The easiest way is to construct surrogate models in advance for time-consuming objective functions and constraints. Then, a heuristic optimization algorithm (such as a genetic algorithm (GA) or particle swarm optimization) is used to solve the optimization problem based on the surrogate models (Jin 2011; Tang et al. 2013). During this process, each surrogate model remains fixed and acts as a fast simulator for the corresponding objective function or constraint. The accuracy of the final optimal solution depends on the accuracy of the constructed surrogate models. Since the optimal solution to the original problem is not known in advance, the designer must ensure that each surrogate model has a high prediction accuracy throughout the entire design space to ensure that the corresponding optimal solution will be close to the true optimal solution to the original problem. Consequently, surrogate model construction requires the calculation of a large number of sample points.

In addition, there is another way to construct surrogate models by means of surrogate-assisted meta-heuristic algorithms (Jin et al. 2002; Jin 2005; Le et al. 2013; Regis 2013; Mlakar et al. 2015; Akhtar and Shoemaker 2016), which dynamically update the surrogate models throughout their evolutionary processes. The surrogate models in surrogate-assisted meta-heuristic algorithms are used for two main purposes: (1) to guide the generation of the initial population and to

P. Jiang et al., *Surrogate Model-Based Engineering Design and Optimization*,
Springer Tracts in Mechanical Engineering,
https://doi.org/10.1007/978-981-15-0731-1_7

perform cross-mutation operations and (2) to approximate the fitness values of some individuals or populations. During the evolutionary process, the surrogate models are updated with the true responses of the new individuals. In surrogate-assisted meta-heuristic algorithms, surrogate models are used only as auxiliary tools for prediction and approximation during the evolutionary process. The process of a surrogate-assisted meta-heuristic algorithm is still based on the framework of traditional meta-heuristic algorithms, and optimization is carried out through the evolution of the population. Compared with traditional meta-heuristic algorithms, surrogate models can reduce the need for time-consuming simulations. However, since the optimization process is still based on the evolution of population, there is a limit on the reduction in the time consumption for simulations that can be achieved by using surrogate models. That is, a surrogate-assisted meta-heuristic algorithm still requires large numbers of calculations of the objective function and constraints to obtain a reliable optimal solution.

The third way to construct surrogate models is to use a surrogate-based optimization (SBO) algorithm (Schonlau 1997; Jones et al. 1998; Gary Wang et al. 2001; Jones 2001; Wang 2003; Wang et al. 2004; Regis and Shoemaker 2005; Shan and Wang 2005; Regis and Shoemaker 2007; Forrester et al. 2008; Sharif et al. 2008; Regis and Shoemaker 2013; Yondo et al. 2018). Such algorithms are completely different from meta-heuristic algorithms. They rely only on the information provided by the surrogate model to explore and optimize the original problem by continuously supplementing and updating the sample points. An SBO algorithm is very cautious in selecting each updated point. The information of the surrogate model is usually fully explored and analysed to balance the relationship between the global search (adding updated sample points in areas with few sample points) and the local search (adding updated sample points near the current optimal solution). Finally, the point with the most potential is selected as the next sample point. Because SBO algorithms can make full use of the information of the surrogate model and fully exploit the potential for improvement provided by each new sample point, they often require fewer calculations of the objective function and constraints than surrogate-assisted meta-heuristic algorithms to find the optimal solution of the original problem. Consequently, scholars have shown great interest in research on SBO algorithms: the types of surrogate models used, the infill criteria for selecting updated sample points, the algorithm procedures, etc. The general flowchart of an SBO algorithm is shown in Fig. 7.1. A small number of initial sample points are generated to construct the initial surrogate model at first. New sample points are selected based on certain specific infill criteria to update the surrogate model until the iterative process is terminated. The number of sample points and the surrogate model itself are constantly changing throughout the iterative process.

In SBO algorithms, the selection of the updated sample points directly affects the final optimization results and computational efficiency. The existing researches indicate that the updated sample points can be selected based on several commonly used infill criteria, which can be divided into three main categories (Sóbester et al. 2005; Ponweiser et al. 2008; Liu et al. 2012; Parr et al. 2012):

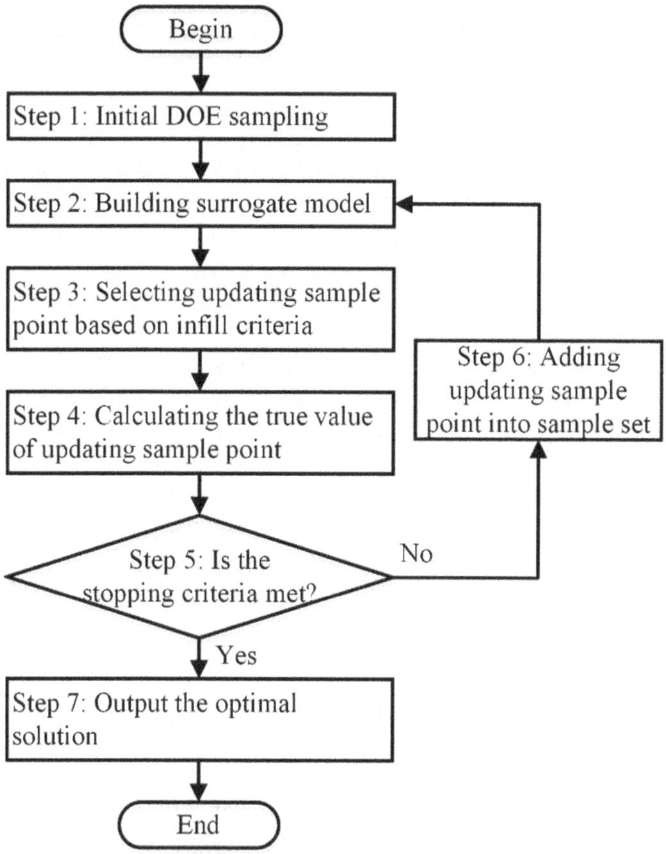

Fig. 7.1 The flowchart of the SBO algorithm

(1) Local infill criteria:

The most commonly used local infill criterion is to minimize the response surface criterion. The local exploitation with such a criterion is excellent, and the speed of convergence is fast, but the optimization process can easily fall into a locally optimal solution. However, this criterion can be adapted for application to any surrogate model.

(2) Global infill criteria:

The global exploration performance of global infill criteria is good. However, their ability to drive local exploitation is weak. The accuracy of the final optimal solution cannot be guaranteed, and the overall number of calculations is large.

(3) Infill criteria balancing global exploration and local exploitation:

Such criteria can be regarded as attempting to fill in the gaps between existing sample points to build a surrogate model that is accurate throughout the design space. For example, a kriging model yields not only the prediction response value at a design point but also the corresponding prediction variance. The prediction variance can be used to represent the uncertainty of the prediction of the kriging model at this design point. It also describes the sparseness of the sample points in the neighbourhood where the design point is located. The larger the prediction variance is, the fewer sample points there are in the neighbourhood of the current design point, and the more inaccurate the kriging model is for predictions in this region. Thus, the number of sample points can be appropriately increased in this region to realize more effective exploration of the region with the kriging model. Since kriging models can provide both prediction response values and variances at the design points, scholars have proposed a variety of infill criteria to balance the relationship between global exploration and local exploitation.

In this chapter, three different kinds of commonly used infill criteria balancing the relationship between global exploration and local exploitation are introduced: the expected improvement criterion (Alexandrov et al. 1998; Sasena 2002; Huang et al. 2006b; Forrester et al. 2006; Kleijnen et al. 2012; Xiao et al. 2012, 2013; Chaudhuri and Haftka 2014), the probability of improvement (PI) criterion (Keane 2006; Viana and Haftka 2010; Couckuyt et al. 2014) and the lower confidence bound (LCB) criterion (Laurenceau et al. 2010; Srinivas et al. 2012; Desautels et al. 2014; Zheng et al. 2016). In addition, some improved forms of these infill criteria that can be used to solve constrained optimization problems are also demonstrated (Audet et al. 2000; Sasena et al. 2002; Basudhar et al. 2012). We will use a kriging model as an illustration. Figure 7.2 shows the kriging model for the following one-dimensional function (Forrester et al. 2008): $f(x) = (6x - 2)^2 \sin(12x - 4)$. The sample points are selected to be [0.0, 0.333, 0.667, 1.0]. The solid red line represents the true function. The four sample points are denoted by dots. The

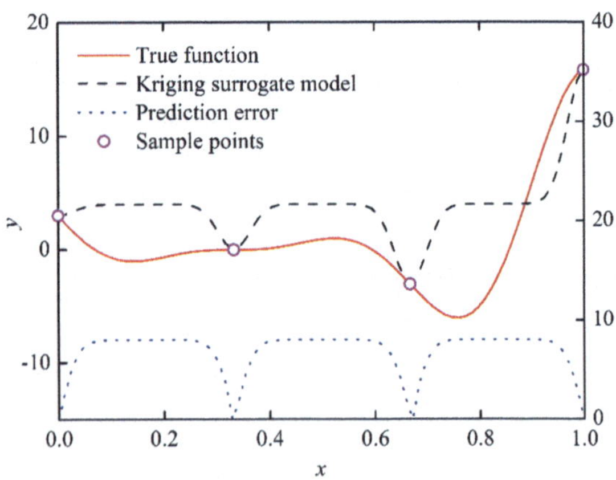

Fig. 7.2 Kriging model for a one-dimensional function

dashed black line represents the response values predicted by the kriging model, and the blue dotted line represents the prediction error. In Fig. 7.2, the left axis is used to represent the values of the true function, the sample points and the responses predicted by the kriging model, while the right axis shows the prediction error of the kriging model. The kriging predictor accurately passes through each sample point, and the error at each sample point is zero, indicating that the kriging model is an interpolation model. The kriging predictor provides not only the predicted response value $\hat{y}(\mathbf{x})$ at any point but also the corresponding prediction error $s(\mathbf{x})$. By virtue of these beneficial characteristics of kriging models, the trusted region for each prediction response value is known. Additionally, kriging models can be widely used in surrogate-model-assisted design optimization.

7.1.1 Expected Improvement Criteria

The expected improvement (EI) criterion was first proposed by Schonlau and Jones (Schonlau 1997; Jones et al. 1998) for the sequence update algorithm. Additionally, the constrained expected improvement (CEI) criterion introduced in Schonlau's doctoral dissertation (Schonlau 1997) can be used to solve constrained optimization problems. The EI criterion is one of the most widely studied infill criteria for kriging models that are used to approximate time-consuming objective functions and constraint functions. The prediction response values and errors provided by a kriging model are useful in determining the update criteria (the EI criterion and the CEI criterion).

7.1.1.1 The Unconstrained Expected Improvement (EI) Criterion

The EI criterion is mainly used to solve time-consuming unconstrained optimization problems. Consider the following unconstrained optimization problem:

$$\begin{aligned} \min \quad & f(\mathbf{x}) \\ s.t. \quad & x_i^l \le x_i \le x_i^u,\ i = 1, 2, \ldots, n \\ & \mathbf{x} = [x_1, x_2, \ldots x_n] \end{aligned} \tag{7.1}$$

where n is the number of design variables and x_i^l and x_i^u are the lower and upper bounds, respectively, on design variable x_i, where $x_i^l < x_i^u$. Since a meta-heuristic algorithm would require thousands of calculations of the objective function, such an algorithm cannot be directly used to solve the original problem. Therefore, a kriging model is used to approximate the objective function of the original problem. Based on the kriging prediction and the error function, the EI criterion is constructed. In each iteration, the next sample point is selected in accordance with the EI criterion and the design space is searched by considering the new sample point, and finally, the optimal solution to the original problem is found.

The EI criterion is essential since it determines the selection of each new sample point and the search direction of the optimization process. For any unobserved point $\mathbf{x}$, the kriging model provides a prediction response value $\hat{y}(\mathbf{x})$ and its standard deviation $s(\mathbf{x})$. The main problem in designing the optimization process is determining how to use these two pieces of information provided by the kriging model to select the most promising point as the next sample point. One possibility is to select the point with the minimum predictive response value $\hat{y}(\mathbf{x})$ as the next sample point; in this way, the area near the current optimal solution can be fully explored to further improve the current optimal solution. However, this kind of search will remain concentrated in a local area, which may cause the search process to become trapped near a certain locally optimal solution to the original problem. Another possibility is to select the point with the maximum standard deviation $s(\mathbf{x})$ as the next sample point; in this way, unknown areas can be explored to the greatest possible extent, and the next sample point will be selected in an area where the sample points are sparse, thus allowing the search process to escape from the current local region. However, this kind of search is very slow, and a large number of supplementary sample points will be required to find the optimal solution to the original problem.

The EI criterion balances and synthesizes these two search modes (local and global search). Any unobserved point $\mathbf{x}$ can be regarded as a random variable with a normal distribution with a mean of $\hat{y}(\mathbf{x})$ and a standard deviation of $s(\mathbf{x})$:

$$Y(\mathbf{x}) \sim N(\hat{y}(\mathbf{x}), s(\mathbf{x})) \tag{7.2}$$

Suppose that the minimum value of the objective function among the current sample points is $f_{\min}$, which is the current optimal solution. Then, the improvement with respect to the current optimal solution that is achieved at the unobserved point $\mathbf{x}$ can also be regarded as a random variable:

$$\mathrm{I}(\mathbf{x}) = \max(f_{\min} - Y(\mathbf{x}), 0) \tag{7.3}$$

In probability theory, the expected value reflects the average value of a random variable. To evaluate the improvement, we can take the expected value of the corresponding variable as follows:

$$\begin{aligned} \mathrm{EI}(\mathbf{x}) &= \mathrm{E}[\max(f_{\min} - Y(\mathbf{x}), 0)] \\ &= \int_{-\infty}^{f_{\min}} (f_{\min} - Y) \frac{1}{\sqrt{2\pi} s(\mathbf{x})} \exp\left(-\frac{\hat{y}(\mathbf{x})^2}{2s(\mathbf{x})^2}\right) dY \end{aligned} \tag{7.4}$$

By solving Eq. (7.4), the following expression for the EI function can be obtained:

$$\mathrm{EI}(\mathbf{x}) = (f_{\min} - \hat{y}(\mathbf{x}))\Phi\left(\frac{f_{\min} - \hat{y}(\mathbf{x})}{s(\mathbf{x})}\right) + s(\mathbf{x})\phi\left(\frac{f_{\min} - \hat{y}(\mathbf{x})}{s(\mathbf{x})}\right) \quad (7.5)$$

where $\phi(\cdot)$ and $\Phi(\cdot)$ represent the probability density function and the cumulative distribution function, respectively, of the standard normal distribution. It can be seen from Eq. (7.5) that the EI function is a nonlinear combination of $\hat{y}(\mathbf{x})$ and $s(\mathbf{x})$. The first term of the EI function increases as $\hat{y}(\mathbf{x})$ decreases, so it tends to select a point with a smaller predicted response value as the next sample point. The second term of the EI function increases as $s(\mathbf{x})$ increases, so it tends to select a point with a larger prediction variance as the next sample point. The EI function thus considers both the smallest value of the mean and the region with the largest standard deviation, thereby automatically balancing the exploration and development capabilities of the optimization process by considering both local and global search criteria.

Figure 7.3 shows the EI function for the previously introduced one-dimensional function. The solid line represents the true function, the dots represent the four sample points, the dashed line represents the kriging model built for this function, and the dotted line represents the EI function. The left axis shows the values of the true function, the sample points and the kriging model, while the EI function is represented on the right axis. In, the EI function is a continuous function. The values of the EI function at the sample points are 0, while the values at unobserved points are greater than 0.

By taking the partial derivatives of $\mathrm{EI}(\mathbf{x})$ with respect to $\hat{y}(\mathbf{x})$ and $s(\mathbf{x})$, respectively, we obtain

$$\frac{\partial \mathrm{EI}(\mathbf{x})}{\partial \hat{y}(\mathbf{x})} = -\Phi\left(\frac{f_{\min} - \hat{y}(\mathbf{x})}{s(\mathbf{x})}\right) < 0 \quad (7.6)$$

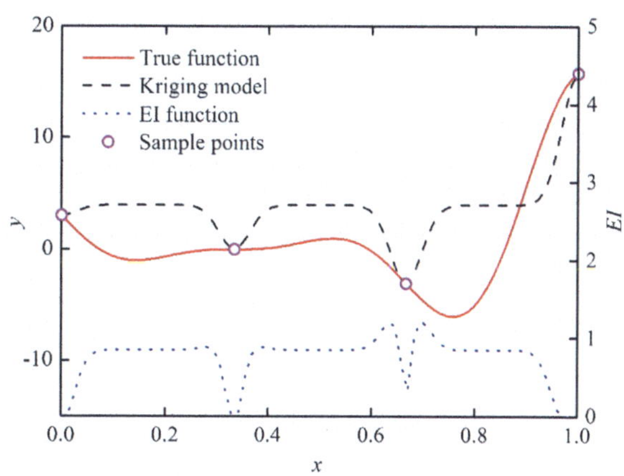

Fig. 7.3 EI function for the one-dimensional function

and

$$\frac{\partial \mathrm{EI}(\mathbf{x})}{\partial s(\mathbf{x})} = \phi\left(\frac{f_{\min} - \hat{y}(\mathbf{x})}{s(\mathbf{x})}\right) > 0 \tag{7.7}$$

Equations (7.6) and (7.7) show that the EI function is monotonically decreasing in $\hat{y}(\mathbf{x})$ and monotonically increasing in $s(\mathbf{x})$. Accordingly, two characteristics of the EI function can be derived as follows. For any two unobserved points $\mathbf{x}^{(1)}$ and $\mathbf{x}^{(2)}$, the response values predicted by the kriging model are $\hat{y}(\mathbf{x}^{(1)})$ and $\hat{y}(\mathbf{x}^{(2)})$, respectively, and the predicted standard deviations of the kriging model are $s(\mathbf{x}^{(1)})$ and $s(\mathbf{x}^{(2)})$, respectively. With this notation, the following characteristics of the EI function can be described:

(1) If $s(\mathbf{x}^{(1)}) = s(\mathbf{x}^{(2)})$ and $\hat{y}(\mathbf{x}^{(1)}) < \hat{y}(\mathbf{x}^{(2)})$ are satisfied, then $\mathrm{EI}(\mathbf{x}^{(1)}) > \mathrm{EI}(\mathbf{x}^{(2)})$.
(2) If $s(\mathbf{x}^{(1)}) > s(\mathbf{x}^{(2)})$ and $\hat{y}(\mathbf{x}^{(1)}) = \hat{y}(\mathbf{x}^{(2)})$ are satisfied, then $\mathrm{EI}(\mathbf{x}^{(1)}) > \mathrm{EI}(\mathbf{x}^{(2)})$.

In other words, when the next sample point is selected based on the EI criterion, if the standard deviations of the predictions at two points are the same, then the point with the smaller predicted response value will be selected. Similarly, if two points have the same predicted response value, then the point with the larger standard deviation will be selected. This property is called the monotonicity of the EI criterion.

Since a kriging model is an interpolation model, the standard deviations of the model predictions at all sample points are zero, i.e. $s(\mathbf{x}^{(i)}) = 0, i = 1, 2, \ldots, N$. From Eq. (7.5), we know that when $s(\mathbf{x}) = 0$, $\mathrm{EI}(\mathbf{x}) = 0$. That is, the values of the EI function at the sample points are equal to zero. If $\mathbf{x}$ is an unobserved point, then the standard deviation of the prediction of the kriging model at this point is greater than zero, i.e. $s(\mathbf{x}) > 0$. From Eq. (7.5), we know that when $s(\mathbf{x}) > 0$, $\mathrm{EI}(\mathbf{x}) > 0$; this means that the values of the EI function at unobserved points are greater than zero. Therefore, when the point with the maximum value of the EI function is selected as the next sample point in each iteration, the selected points are guaranteed not to be sample points (because the values of the EI function at all sample points are zero, whereas the values at all unobserved points are greater than zero). Thus, the new sample points selected based on the EI function must be different from the previously selected sample points. According to the convergence criteria of Torn and Zilinskas (Törn and Zilinskas 1989), the selection process based on the EI function will drive the optimization process to converge to the optimal solution to the original problem. This property is called the convergence of the EI criterion.

7.1.1.2 Procedures for the SBO Algorithm Driven by the EI Criterion

The procedures for the SBO algorithm driven by the EI criterion are shown in Algorithm 7.1. This is a typical two-step SBO algorithm: the first step is to apply a design of experiments method to generate the initial sample points, and the second step is to iterate the loop until the stopping criterion is satisfied. Each iteration of the SBO algorithm consists of three steps: first, constructing the kriging model based on the set of sample points, as shown in line 2 of Algorithm 7.1; second, selecting the point with the largest value of the EI function as the new sample point, as shown in line 3; and third, calculating the true response value at the new sample point and adding this new point into the set of sample points, as shown in lines 4–8. The SBO algorithm finds the optimal solution by continuously selecting and calculating new sample points. It does not consider the overall accuracy of the kriging model. Instead, the kriging model is used to assist in the optimization procedure. After all, finding the optimal solution to the original problem is the ultimate goal of the SBO algorithm.

Algorithm 7.1 Procedures for the SBO algorithm driven by the EI criterion

Require: initial sample points $(\mathbf{X}, \mathbf{y})$

Find: the optimal solution $(\mathbf{x}_{\min}, y_{\min})$

1. **while** the stopping criterion is not satisfied **do**
2. Construct a kriging model based on the set of sample points $(\mathbf{X}, \mathbf{y})$
3. $x^{(u)} = \arg\max \ \mathrm{EI}(\mathbf{x})$
4. Calculate the true response value $y(\mathbf{x}^{(u)})$ at the new sample point $\mathbf{x}^{(u)}$
5. $\mathbf{X} \leftarrow \mathbf{X} \cup \mathbf{x}^{(u)}$
6. $\mathbf{y} \leftarrow \mathbf{y} \cup y(\mathbf{x}^{(u)})$
7. $y_{\min} \leftarrow \min(y)$
8. $x_{\min} \leftarrow \mathbf{x} \in \mathbf{X}; y(\mathbf{x}) = y_{\min}$
9. **end while**

Figure 7.4 shows the process of searching based on the EI criterion for the one-dimensional function considered as an example here. Each row represents one round of iteration. The first row represents the initial state, and the optimal solution to the one-dimensional function is found through six iterations. In the left column, the red solid lines represent the true one-dimensional function, the black dashed lines represent the constructed kriging models, the blue dots indicate the existing sample points, and the pink dots denote the newly added sample points. In the right column, the black solid lines represent the EI function, and the pink dots are the points with the maximum value of the EI function, that is, the new sample points selected based on the EI criterion. The blue dots in Fig. 7.4a are the four initial sample points, and the dotted line represents the initial kriging model. After the

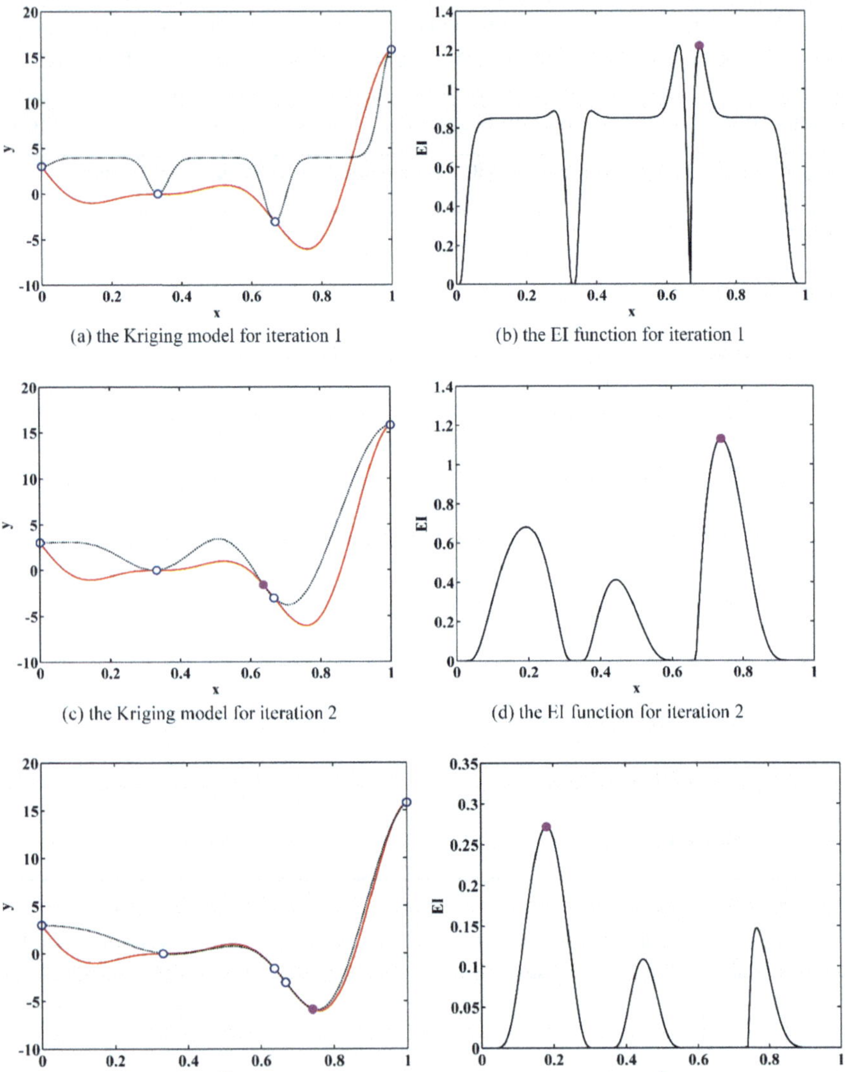

(a) the Kriging model for iteration 1

(b) the EI function for iteration 1

(c) the Kriging model for iteration 2

(d) the EI function for iteration 2

(e) the Kriging model for iteration 3

(f) the EI function for iteration 3

Fig. 7.4 The iterative process driven by the EI criterion for the one-dimensional function

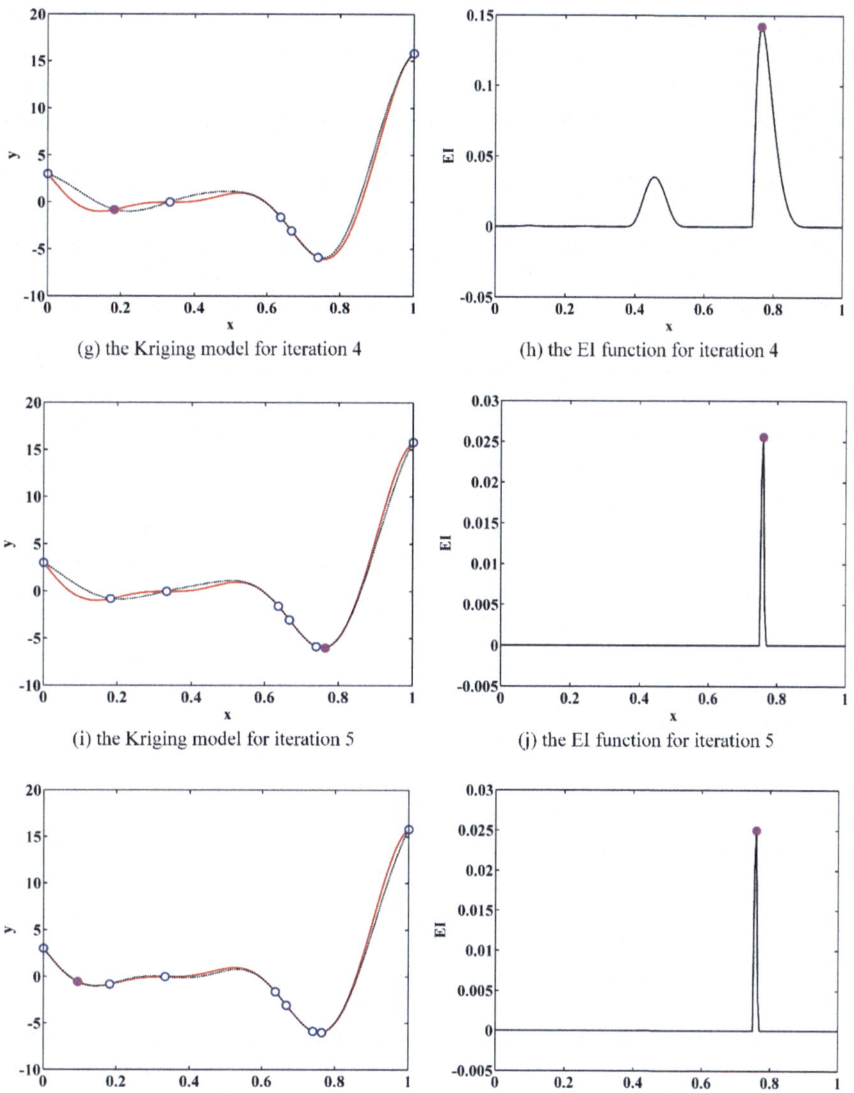

(g) the Kriging model for iteration 4

(h) the EI function for iteration 4

(i) the Kriging model for iteration 5

(j) the EI function for iteration 5

(k) the Kriging model for iteration 6

(l) the EI function for iteration 6

Fig. 7.4 (continued)

initial kriging model is constructed, the EI function of the initial kriging model (the black solid line in Fig. 7.4b) is used to identify the point with the maximum value of the EI function (the pink dot in Fig. 7.4b), which is chosen as the next sample point. Then, the true response value at the chosen sample point is calculated, and this point is added to the set of sample points. Then, the second round of iteration begins. The kriging model is reconstructed (the black dashed line in Fig. 7.4c) based on the updated set of sample points including the newly added sample point, and the new EI function is obtained (the black solid line in Fig. 7.4d). The second sample point is selected as the one that maximizes the EI function (the pink dot in Fig. 7.4d). The SBO algorithm iteratively searches the design space in this way and finally finds the optimal solution to the one-dimensional function after six iterations.

This SBO algorithm is quite different from derivative-based or meta-heuristic optimization algorithms. The SBO algorithm does not search based on a specific optimization path, as a derivative-based optimization algorithm does. Instead, the design space is searched under the guidance of the EI function. There is no correlation between two sequential search points (newly added sample points), and the search process often jumps from one region to another, thus preventing it from falling into a local optimum. In addition, the SBO algorithm does not search based on the evolution of a population, as a meta-heuristic algorithm does, but rather searches for the optimal solution to the original problem by finding and adding a single point in each iteration. There are no inheritance or mutation relationships between the search points in two successive iterations; instead, the search points are strictly selected on the basis of the maximum value of the EI function. Derivative-based optimization algorithms often use only the derivative information from the current local region, while meta-heuristic optimization algorithms often use only the information of the current population. By contrast, each selection process in the SBO algorithm makes full use of the information from each sample point. Based on all of the sample points, a kriging model is constructed to approximate the original problem, and the available information about the original problem is optimized based on the EI function.

In addition, the EI function is an analytical expression, and it is easily and quickly calculated. In each iteration of the SBO algorithm, the EI function needs to be optimized to select the next sample point. Since the EI function is usually a highly nonlinear function with multiple peaks, it is often necessary to use a meta-heuristic optimization algorithm to find its maximum value, which requires thousands of calculations of the EI function. However, since the EI function is easily and quickly calculated, the whole optimization process takes only a few seconds to a few minutes. Thus, compared to time-consuming simulations, the optimization of the EI function requires a negligible amount of time.

In general, an SBO algorithm is suitable for solving time-consuming optimization problems. Compared with derivative-based optimization methods, an SBO algorithm has better global search capabilities and does not require the derivative information of the original problem. In addition, compared with meta-heuristic optimization algorithms, an SBO algorithm requires only a small number of calculations of the objective function, which can greatly reduce the optimization time and improve the efficiency of optimization. In fact, an SBO algorithm is a general optimization method and can also be used to solve non-time-consuming optimization problems. For non-time-consuming optimization problems, however, the greatest advantage of the SBO algorithm (requiring only a small number of calculations of the objective function) provides no obvious benefit, and it may be more appropriate to use a meta-heuristic algorithm.

7.1.1.3 The Constrained Expected Improvement (CEI) Criterion

The EI criterion is mainly used for unconstrained optimization problems. However, most practical optimization problems contain one or more constraints. A single-objective optimization problem with constraints can be described as follows:

$$\begin{aligned} \min \quad & f(\mathbf{x}) \\ s.t. \quad & g_j(\mathbf{x}) \leq 0,\ j = 1, 2, \ldots, c \\ & x_i^l \leq x_i \leq x_i^u,\ i = 1, 2, \ldots, n \\ & \mathbf{x} = [x_1, x_2, \ldots x_n] \end{aligned} \tag{7.8}$$

where c is the number of constraint functions. In this chapter, it is assumed that all constraint functions can be expressed as inequality functions of the form $g(\mathbf{x}) \leq 0$. For example, an inequality constraint function of the form $g(\mathbf{x}) \geq 0$ can be transformed into $-g(\mathbf{x}) \leq 0$ by multiplying by -1, and an equality constraint function of the form $g(\mathbf{x}) = 0$ can be represented by two inequality constraint functions, $g(\mathbf{x}) \geq -\varepsilon$ and $g(\mathbf{x}) \leq +\varepsilon$, where ε is a small positive number.

It is also assumed that both the objective function $f(\mathbf{x})$ and the constraint functions $g_j(\mathbf{x})$ need to be calculated by means of time-consuming simulations. For constraint functions that are analytical expressions and quick to calculate, they can be directly considered when maximizing the EI function. The optimization problem of selecting the next sample point can be transformed from minimizing the unconstrained EI function to minimizing the constrained EI function:

$$\begin{aligned} \min \quad & \mathrm{EI}(\mathbf{x}) \\ s.t. \quad & g_{ju}(\mathbf{x}) \leq 0,\ ju = 1, 2, \ldots, u \\ & \mathbf{x} = [x_1, x_2, \ldots x_n] \end{aligned} \tag{7.9}$$

where $\mathrm{EI}(\mathbf{x})$ is the unconstrained EI function, g_{ju} represents the ju-th constraint function with an analytical expression, and u denotes the total number of constraint functions with analytical expressions. By contrast, if the calculation of a constraint function involves time-consuming simulations, that constraint function cannot be directly considered. Instead, a kriging model needs to be constructed for the constraint function. Thus, in the optimization process, the number of time-consuming simulations of constraint functions can be significantly reduced. An optimization problem for which both the objective function and one or more constraint functions involve time-consuming simulations can be called a time-consuming constrained optimization problem.

Schonlau proposed the CEI criterion based on the consideration of time-consuming constraint functions. After the initial sample points are generated, it is necessary to calculate both the true response values $\{y^{(1)}, y^{(2)}, \ldots, y^{(N)}\}$ of the objective function at the sample points and the true response values $\left\{g_j^{(1)}, g_j^{(2)}, \ldots, g_j^{(N)}\right\} j = 1, 2, \ldots, c$, of each constraint function at the sample points. When constructing the kriging models, it is necessary to construct a kriging model for both the objective function and each constraint function. For any unobserved point $\mathbf{x}$, the predicted response value of the objective function is $\hat{y}(\mathbf{x})$, and the predicted error is $s(\mathbf{x})$. The predicted response value of the j-th constraint function is $\hat{g}_j(\mathbf{x})$, and the predicted error is $e_j(\mathbf{x})$. That is, the objective function is a random variable that satisfies.

$$Y(\mathbf{x}) \sim N(\hat{y}(\mathbf{x}), s(\mathbf{x})) \tag{7.10}$$

The j-th constraint function is also a random variable and satisfies

$$G_j(\mathbf{x}) \sim N\left(\hat{g}_j(\mathbf{x}), e_j(\mathbf{x})\right), j = 1, 2, \ldots, c \tag{7.11}$$

Schonlau defines an improved function that satisfies all constraint functions as follows:

$$\mathrm{I}(\mathbf{x}) = \begin{cases} \max\left(f_{\min}^* - Y(\mathbf{x}), 0\right), & \text{if } G_j(\mathbf{x}) \leq 0 \quad \text{for } j = 1, 2, \ldots, c \\ 0 & \text{otherwise} \end{cases} \tag{7.12}$$

where $f_{\min}^*$ is the minimum value of the objective function in the current set of sample points that satisfies all constraint functions. The physical meaning of the improved constraint function is that when $\mathbf{x}$ satisfies all constraint functions, the value of the improved function is $\max\left(f_{\min}^* - Y(\mathbf{x}), 0\right)$, and when $\mathbf{x}$ does not satisfy all constraint functions, i.e. at least one constraint function is not satisfied, the value of the improved function is 0. It is assumed that the objective function $Y(\mathbf{x})$ and each constraint function $G_j(\mathbf{x})$ are independent of each other. Thus, the CEI criterion is formulated as

$$\begin{aligned} \mathrm{CEI}(\mathbf{x}) &= \mathrm{EI}(\mathbf{x}) \times \prod_{j=1}^{c} \Pr\left(G_j(\mathbf{x}) \leq 0\right) \\ &= \left[(f_{\min} - \hat{y}(\mathbf{x}))\Phi\left(\frac{f_{\min} - \hat{y}(\mathbf{x})}{s(\mathbf{x})}\right) + s(\mathbf{x})\phi\left(\frac{f_{\min} - \hat{y}(\mathbf{x})}{s(\mathbf{x})}\right)\right] \times \prod_{j=1}^{c} \Phi\left(\frac{-\hat{g}_j(\mathbf{x})}{e_j(\mathbf{x})}\right) \end{aligned} \tag{7.13}$$

where $\text{EI}(\mathbf{x})$ is the EI function without considering the constraint functions and $\Pr(G_j(\mathbf{x}) \leq 0)$ is the probability that the point $\mathbf{x}$ satisfies the j-th constraint function. Thus, $\prod_{j=1}^{c} \Pr(G_j(\mathbf{x}) \leq 0)$ is the probability that the point $\mathbf{x}$ satisfies all constraint functions. Usually, the probability that the point $\mathbf{x}$ satisfies all constraint functions is called the probability of feasibility (PoF):

$$\text{PoF}(\mathbf{x}) = \prod_{j=1}^{c} \Pr(G_j(\mathbf{x}) \leq 0) = \prod_{j=1}^{c} \Phi\left(\frac{-\hat{g}_j(\mathbf{x})}{e_j(\mathbf{x})}\right) \quad (7.14)$$

The first factor of the CEI criterion (in Eq. (7.13)) tends to select a point where the expected improvement is large, while the second factor tends to select a point where the PoF is large. By considering the product of these two factors, the CEI criterion selects a point that is likely to satisfy all constraint functions and also has a large expected improvement as the next sample point. To demonstrate the optimization properties of the CEI criterion, the Branin optimization problem is revised by adding only one simple constraint (Forrester et al. 2008):

$$\begin{aligned} f &= \left(x_2 - \frac{5.1}{4\pi^2}x_2 + \frac{5}{\pi}x_1 - 6\right)^2 + 10\left[\left(1 - \frac{1}{8\pi}\right)\cos x_1 + 1\right] + 5x_1, \\ g &= -x_1x_2 + 20 \leq 0 \\ x_1 &\in [-5, 10], x_2 \in [0, 15] \end{aligned} \quad (7.15)$$

Figure 7.5a depicts the constrained Branin optimization problem, where the solid line represents the contour of the objective function. The boundary represented by the blue solid line indicates where the constraint function is equal to zero. The lower left area is the infeasible region, while the upper right area represents the feasible region. The blue star denotes the globally optimal feasible solution to the constrained Branin optimization problem, which is located on the boundary of the constraint function.

Figure 7.5b–d shows the contours of the EI function, the PoF function and the CEI function for the constrained Branin optimization problem. The more darkly an area is coloured, the larger the function value in that area. The white dots represent the sample points used to build the kriging model. In Fig. 7.5b, the positions of the three peaks of the EI function are very close to the positions of the three minimum values of the objective function of the constrained Branin optimization problem, which means that the EI function can best search the objective space. As shown in Fig. 7.5c, the PoF function can accurately approximate the constraint function. The value of the PoF function in the lower left infeasible region is 0, while the value in the upper right feasible region is 1. Additionally, near the constraint boundary between the lower left and the upper right, the value of the PoF function changes

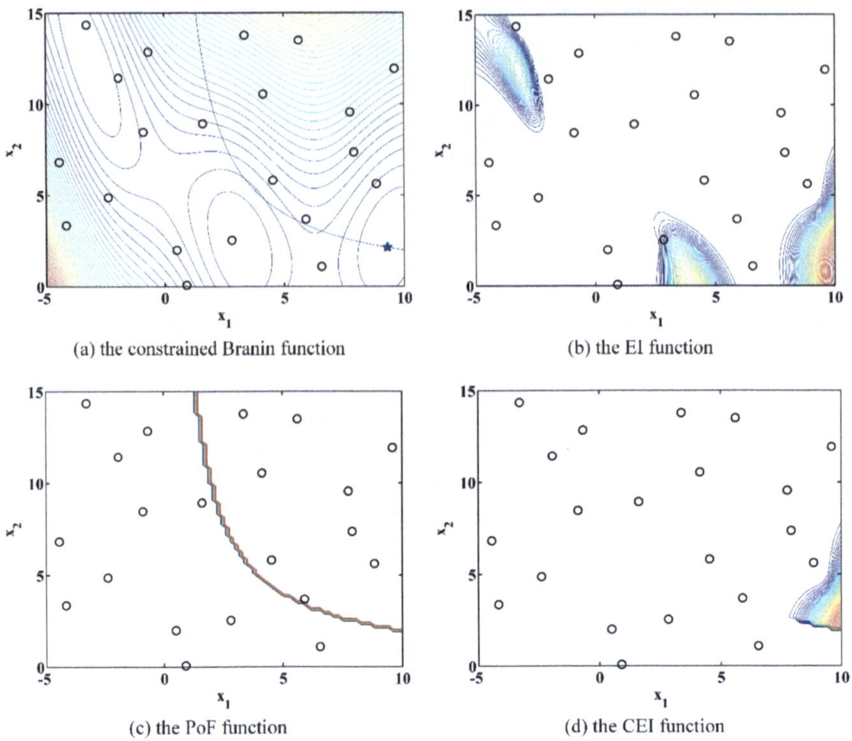

Fig. 7.5 The constrained Branin problem and its corresponding EI, PoF and CEI functions

sharply from 0 to 1. The CEI function shown in Fig. 7.5d is the product of the EI function and the PoF function, meaning that it considers both the objective function and the constraint function. In Fig. 7.5, it can be seen that the peak area on the right of the CEI function is very close to the true optimal feasible solution to the constrained Branin problem, which means that the CEI criterion can be effectively used to search for the optimal solution to the constrained Branin problem.

The CEI criterion is the main criterion used for constrained optimization problems. However, when there is no feasible solution in the set of sample points, the constrained SBO algorithm uses the PoF criterion as the update criterion to select a new sample point that satisfies all constraint functions. Once a feasible solution is found, the constrained SBO algorithm again uses the CEI criterion as the update criterion to improve the current optimal feasible solution.

7.1.1.4 Procedures for the SBO Algorithm Driven by the CEI Criterion

The process of the SBO algorithm driven by the CEI criterion is similar to that of the SBO algorithm driven by the EI criterion. The algorithm can again be divided into two steps: the first step is to apply a design of experiments method to generate the initial sample points, and the second step is to iterate the loop while selecting a new sample point in each iteration. However, the iteration process of the constrained SBO algorithm is slightly different from that of the unconstrained SBO algorithm. The constrained SBO algorithm uses either the PoF criterion or the CEI criterion to select the next point based on whether the current set of sample points contains a feasible solution. If there is no feasible solution among the current sample points, then the constrained SBO algorithm must find a feasible solution; therefore, the update criterion is to maximize the PoF function to select the updated sample point that is most likely to satisfy all constraint functions. If there is at least one feasible solution among the sample points (the initial sample points plus any sample points obtained by maximizing the PoF criterion), then the highest priority of the constrained SBO algorithm is to improve the current optimal feasible solution; therefore, the update criterion is to maximize the CEI function to select the updated sample point that offers the greatest potential improvement to the current optimal feasible solution.

The optimization process of the constrained SBO algorithm driven by the CEI criterion is shown in Algorithm 7.2. First, for the initial sample points, not only the true response values of the objective function but also those of all constraint functions need to be calculated. Similarly, a kriging model must be built not only for the objective function but also for each constraint function, as shown in lines 2–5. Then, before selecting the new sample point, it is necessary to judge whether a feasible solution exists in the current set of sample points. If there is no feasible solution, then the next sample point is selected based on the PoF criterion, whereas if there is at least one feasible solution among the current sample points at this time, then the next sample point is selected based on the CEI criterion, as shown in lines 6–10. After the next sample point is chosen, the constrained SBO algorithm calculates the true value of the objective function and each constraint function at the new sample point and adds this sample point to the set of sample points, as shown in lines 11–16. At the end of each round of iteration, if there is still no feasible solution in the current sample set, then the message 'Cannot find any feasible solution!' is output, whereas if there is a feasible solution in the sample set at this time, then the current optimal feasible solution is updated, as shown in lines 17–21. Thus, the constrained SBO algorithm completes one round of iteration. If the predetermined stopping criterion is not satisfied at this time, then the SBO algorithm returns to line 2 and proceeds to the next iteration.

Algorithm 7.2 Procedures for the SBO algorithm driven by the CEI criterion

Require: initial sample points $\mathbf{X}$, objective function $\mathbf{y}$ and constraint functions $\{\mathbf{g}_1, \mathbf{g}_2, \ldots, \mathbf{g}_c\}$

Find: the optimal solution $(\mathbf{x}_{\min}, y_{\min})$

1. **while** the stopping criterion is not satisfied **do**
2. Construct a kriging model for the objective function based on the set of sample points $(\mathbf{X}, \mathbf{y})$
3. **for** $i = 1$ to c **do**
4. Construct the i-th kriging model for the i-th constraint function based on the set of sample points $(\mathbf{X}, \mathbf{g}_i)$ $(i = 1, 2, \ldots, c)$
5. **end for**
6. **if** no feasible solution exists among the current sample points **then**
7. $x^{(u)} = \arg\max \ \mathrm{PoF}(\mathbf{x})$
8. **else**
9. $x^{(u)} = \arg\max \ \mathrm{CEI}(\mathbf{x})$
10. **end**
11. Calculate the true response values of the objective function and each constraint function at the new sample point $\mathbf{x}^{(u)}$
12. $\mathbf{X} \leftarrow \mathbf{X} \cup \mathbf{x}^{(u)}$
13. $\mathbf{y} \leftarrow \mathbf{y} \cup y(\mathbf{x}^{(u)})$
14. **for** $i = 1$ to c **do**
15. $\mathbf{g}_i \leftarrow \mathbf{g}_i \cup g_i(\mathbf{x}^{(u)})$
16. **end for**
17. **if** no feasible solution exists among the current sample points **then**
18. Output "Cannot find any feasible solution!"
19. **else**
20. Update the current optimal solution to $(x_{\min}, y_{\min})$
21. **end if**
22. **end while**

Figure 7.6 shows the iterative process of the SBO algorithm for the constrained Branin problem. In the left column, the solid lines represent the contour maps of the Branin objective function, the blue solid lines represent the constraint function boundaries, the upper right areas represent the feasible regions, the white dots represent the sample points, the red dots represent the newly added sample points selected based on the CEI criterion, and the blue stars represent the true optimal feasible solution to the constrained Branin optimization problem. In the right column, the cloud images represent the contour plots of the CEI criterion, the white dots represent the sample points and the red dots represent the positions of the maximum values of the CEI function.

The steps of the constrained SBO algorithm are summarized as follows. First, 21 initial sample points are selected in the design space, as shown by the white dots in Fig. 7.6a. Then, the true response values of the objective function and the constraint function at the initial sample points are calculated, and the initial kriging models of the objective function and the constraint function are constructed based on these true response values. Since feasible solutions exist among the initial sample points, the constrained SBO algorithm uses the CEI criterion to search for the next sample point. Based on the predicted response values and prediction errors provided by the kriging models, the contour plot of the CEI function is obtained, as shown in Fig. 7.6b.

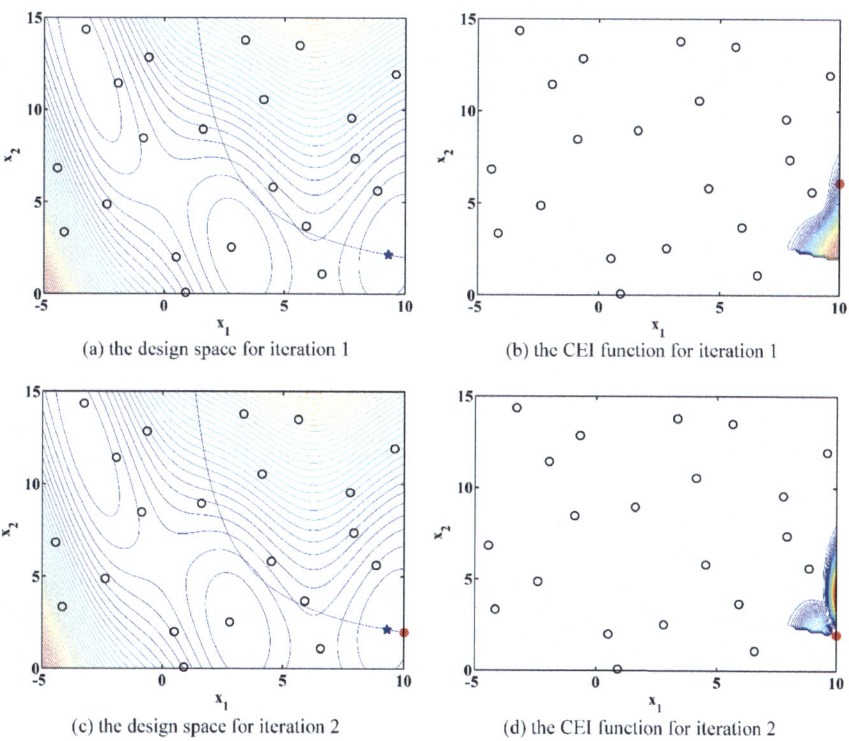

(a) the design space for iteration 1
(b) the CEI function for iteration 1
(c) the design space for iteration 2
(d) the CEI function for iteration 2

Fig. 7.6 The iterative process driven by the CEI criterion for the constrained Branin problem

A meta-heuristic optimization algorithm is applied to obtain the maximum value of the CEI criterion, and the corresponding point is selected as the first new sample point to be added, as shown by the red dot in Fig. 7.6b. After the first new sample point is obtained, the true response values of the objective function and the constraint function at this point are calculated. The first new sample point is added to the sample set, and the kriging models of the objective function and the constraint function are updated. Based on the updated kriging models, the contour plot of the updated CEI function is obtained, as shown in Fig. 7.6d. The second new sample point to be added, as shown by the red dot in Fig. 7.6d, is then also obtained by maximizing the updated CEI function using the meta-heuristic optimization algorithm. As before, the true response values of the objective function and the constraint function at the second new sample point are calculated, and this new sample point is added to the existing sample points. The kriging models of the objective function and the constraint function are updated, and the algorithm proceeds to the next iteration to select the third newly added sample point. In fact, the first newly added sample point is already very close to the real optimal solution to the constrained Branin optimization problem.

In this example, since the constraint function of the Branin optimization problem is very simple, only one iteration of the constrained SBO algorithm needs to be performed based on the CEI criterion to find a solution that is very close to the optimal solution of the constrained Branin problem. When the constraint functions of the problem are more complicated, the constrained SBO algorithm often needs more iterations to find an optimal feasible solution.

It can be seen from the above procedures that the optimization process of the constrained SBO algorithm driven by the CEI criterion is very similar to that of the unconstrained SBO algorithm driven by the EI criterion; the difference is that the unconstrained efficient global optimization (EGO) algorithm uses the EI criterion to dynamically update the kriging model, while the constrained EGO algorithm uses the PoF criterion (when there is no feasible solution in the current sample set) or the CEI criterion (when there is at least one feasible solution in the current sample set) to update multiple kriging models. The PoF and CEI criteria have many properties similar to those of the EI criterion:

(1) Both the PoF criterion and the CEI criterion are functions whose values are greater than or equal to zero. Moreover, when there is no feasible solution in the current sample set, the PoF criterion has a value of zero at every sample point and a value greater than zero at every unobserved point; similarly, when there is at least one feasible solution in the current sample set, the CEI criterion has a value of zero at every sample point and a value greater than zero at every unobserved point.
(2) Both the PoF criterion and the CEI criterion have explicit analytical expressions, which can be quickly and easily calculated. Therefore, a meta-heuristic

optimization algorithm can be used to optimize these criteria to select the next sample point, and the time required to optimize the PoF or CEI function is negligible compared to that of time-consuming simulations.

(3) Since the problem of selecting the next sample point based on the PoF criterion or the CEI criterion is an unconstrained optimization problem, the process of solving it is simpler than that of solving a constrained optimization problem.

7.1.2 Probability of Improvement Criteria

7.1.2.1 The Unconstrained Probability of Improvement (PI) Criterion

For any unobserved point **x** in the design space, when using a kriging model for prediction, the predicted response value at **x** can be regarded as a random variable $Y(\mathbf{x})$ obeying a normal distribution, and its mean $\hat{y}(\mathbf{x})$ and variance $\hat{s}^2(\mathbf{x})$ can be provided by the kriging model. The current optimal solution among the sample points is denoted by $f_{\min}$. The probability of improvement refers to the probability that the predicted response value at an unobserved point will be less than the response value of the current optimal solution $f_{\min}$. The following equation gives the formulation for the PI criterion (Viana and Haftka 2010):

$$\mathrm{P}(\mathbf{x}) = \mathrm{P}\left(\widehat{Y}(\mathbf{x}) < y_{\min}\right) = \Phi\left(\frac{f_{\min} - \hat{y}(\mathbf{x})}{s(\mathbf{x})}\right) \tag{7.16}$$

where $\Phi(\cdot)$ is the standard normal distribution function, $\hat{y}(\mathbf{x})$ is the predicted response value (mean) at the unobserved point and $s(\mathbf{x})$ is the standard deviation of the prediction. When the PI criterion is applied, the point corresponding to the optimal solution that maximizes $\mathrm{P}(\mathbf{x})$ throughout the design space is selected as the next sample point to update the kriging model. Here, the one-dimensional function introduced in Sect. 7.2.1 is used to demonstrate the basic idea of the PI criterion, as shown in Fig. 7.7. The solid red line represents the true function. The black dotted line represents the kriging model constructed based on the sample points (represented by the pink dots in Fig. 7.7). The current optimal solution $f_{\min}$ is the one with the minimum response value among the current sample points (represented by the solid blue line in Fig. 7.7). The green dash-dotted line represents the standard normal probability density function at the unobserved point $x = 0.7$. Thus, the green region in Fig. 7.7 represents the probability of improvement at $x = 0.7$.

7.1.2.2 Procedures for the SBO Algorithm Driven by the PI Criterion

The SBO algorithm driven by the PI criterion follows the same procedures used for the SBO algorithm driven by the EI criterion. The only difference between these

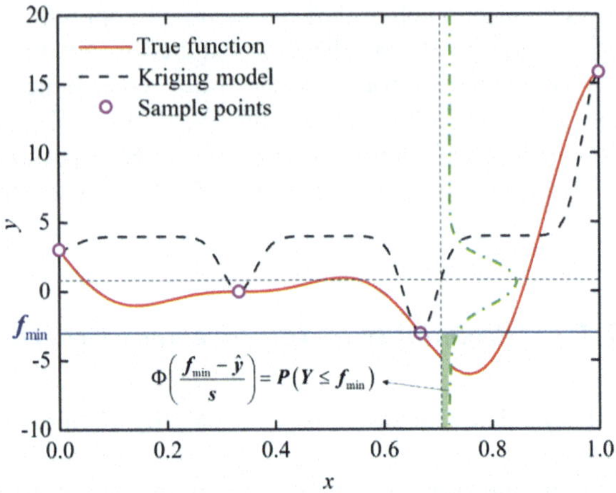

Fig. 7.7 The probability of improvement for the one-dimensional function at $x = 0.7$

procedures is that the SBO algorithm driven by the PI criterion selects the newly added sample points by maximizing the PI function, as shown in Algorithm 7.3, whereas the point with the maximum value of the EI function is selected as the next sample point in the SBO algorithm driven by the EI criterion.

Algorithm 7.3 Procedures for the SBO algorithm driven by the PI criterion

Require: initial sample points $(\mathbf{X},\mathbf{y})$

Find: the optimal solution $(\mathbf{x}_{\min}, y_{\min})$

1. **while** the stopping criterion is not satisfied, **do**
    2. Construct a kriging model based on the set of sample points $(\mathbf{X},\mathbf{y})$
    3. $x^{(u)} = \arg\max \ \mathrm{PI}(\mathbf{x})$
    4. Calculate the true response value $y\left(\mathbf{x}^{(u)}\right)$ at the new sample point $\mathbf{x}^{(u)}$
    5. $\mathbf{X} \leftarrow \mathbf{X} \cup \mathbf{x}^{(u)}$
    6. $\mathbf{y} \leftarrow \mathbf{y} \cup y\left(\mathbf{x}^{(u)}\right)$
    7. $y_{\min} \leftarrow \min(y)$
    8. $x_{\min} \leftarrow \mathbf{x} \in \mathbf{X}; y(\mathbf{x}) = y_{\min}$

    end while

Figure 7.8 shows the iterative process for the one-dimensional function (Forrester et al. 2008) driven by the PI criterion. In the left column, the white dots represent the set of sample points, the pink dots are the new sample points obtained

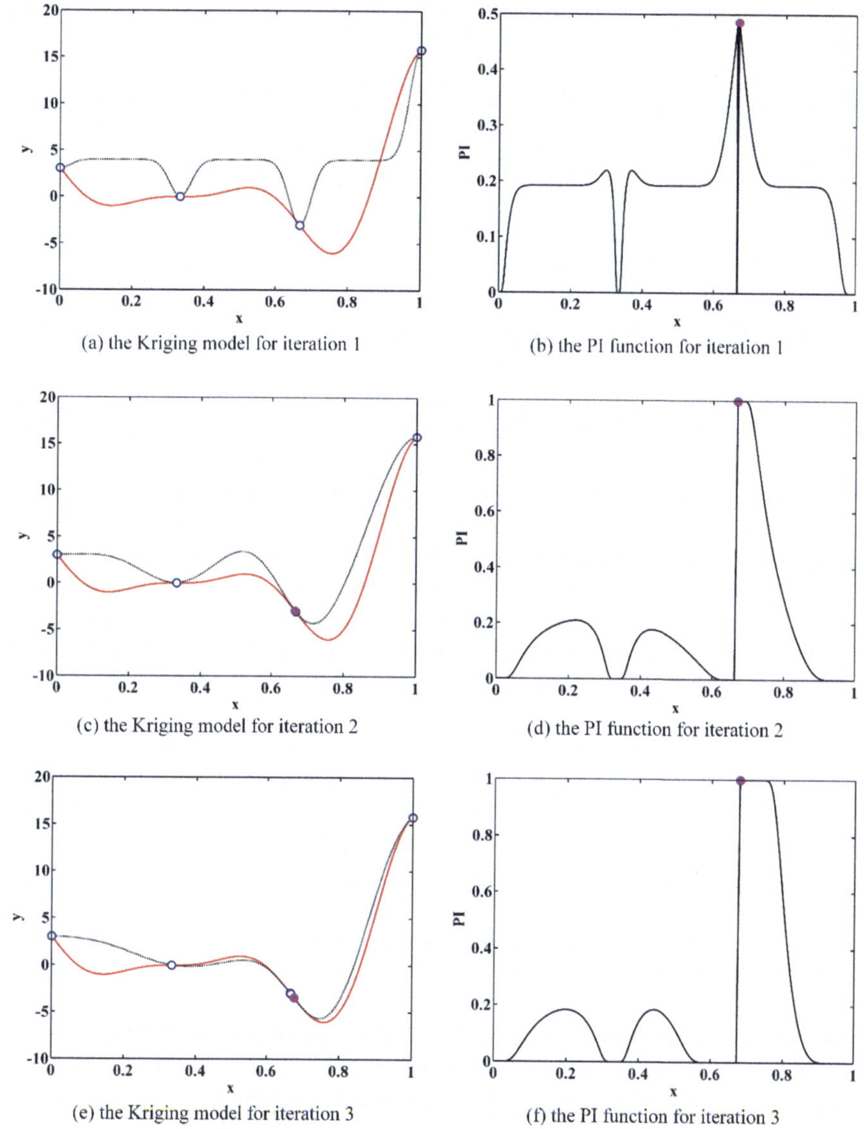

Fig. 7.8 The iterative process driven by the PI criterion for the one-dimensional function

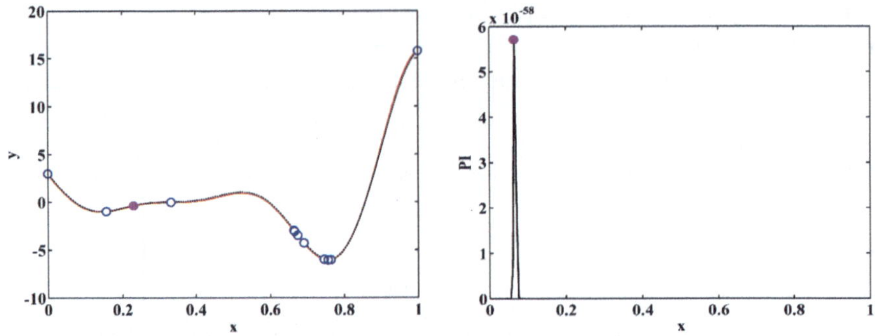

(g) the Kriging model for iteration 9 (h) the PI function for iteration 9

Fig. 7.8 (continued)

by maximizing the PI function, the solid red lines represent the true one-dimensional function and the black dotted lines represent the kriging models constructed based on the sample points. In the right column, the solid black lines represent the PI function, and the pink dots indicate the maximum values of the PI function. The first three rows show the optimization process from iterations 1 to 3, while the last row shows the last round of iteration. As shown in Fig. 7.8a, four initial sample points are used to build the initial kriging model. In the end, eight new sample points have been added to the sample set. As the number of sample points in the neighbourhood of the current optimal solution increases, the predicted standard deviation in that region will decrease. Once the value of the PI function in the neighbourhood of the current optimal solution is sufficiently small, the search will begin to move towards other areas where the predicted standard deviation is large.

In addition, from Fig. 7.8, we can see that the PI function has the same properties as the EI function. The value of the PI function is always greater than or equal to zero. At sample points, the value of the PI function is always equal to 0, and at unobserved points, the value of the PI function is always greater than 0.

7.1.2.3 The Constrained Probability of Improvement (CPI) Criterion

To solve constrained optimization problems, the constrained probability of improvement (CPI) criterion has been proposed. The CPI criterion can be seen as a combination of the PI criterion and the PoF function, which is identical to the PoF function used in the CEI criterion. The CPI criterion is formulated as follows:

$$\begin{aligned} \mathrm{CPI}(\mathbf{x}) &= \mathrm{PI}(\mathbf{x}) \times \prod_{j=1}^{c} \Pr\left(G_j(\mathbf{x}) \le 0\right) \\ &= \Phi\left(\frac{f_{\min} - \hat{y}(\mathbf{x})}{s(\mathbf{x})}\right) \times \prod_{j=1}^{c} \Phi\left(\frac{-\hat{g}_j(\mathbf{x})}{e_j(\mathbf{x})}\right) \end{aligned} \tag{7.17}$$

where $\mathrm{PI}(\mathbf{x})$ is the PI function without considering the constraint functions and $\prod_{j=1}^{c} \Pr\left(G_j(\mathbf{x}) \le 0\right) = \prod_{j=1}^{c} \Phi\left(\frac{-\hat{g}_j(\mathbf{x})}{e_j(\mathbf{x})}\right)$ is the PoF function, which represents the probability that the point $\mathbf{x}$ satisfies all constraint functions.

Since the CPI criterion is similar to the CEI criterion, it has the same capabilities as the CEI criterion. In Fig. 7.9, the constrained Branin optimization problem (Forrester et al. 2008) is used to illustrate the properties of the CEI criterion.

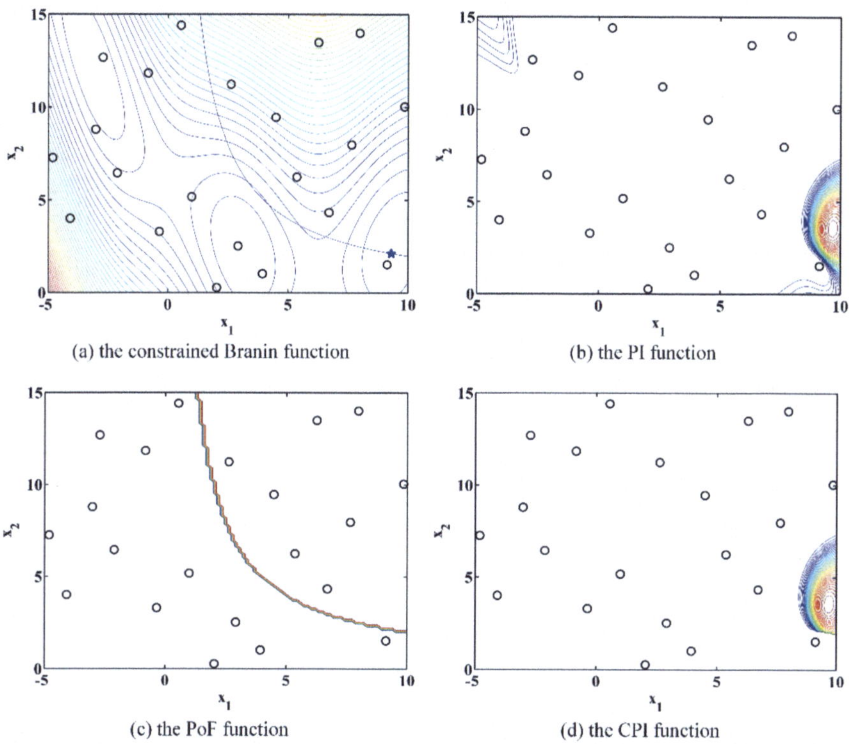

Fig. 7.9 The constrained Branin problem and its corresponding PI, PoF and CPI functions

Figure 7.9a gives the contour plot of the Branin function. The solid blue line represents the constraint function, which divides the design space into feasible and infeasible domains. The lower left area is the infeasible region, while the upper right area is the feasible region. The blue star indicates the true optimal feasible solution to the constrained Branin problem. Figure 7.9b shows the contour plot of the PI function, from which it can be seen that the two peak areas of the PI function are close to two of the three minimum regions of the constrained Branin problem. Figure 7.9c gives the contour plot of the PoF function for the constraint function of the Branin problem, which is consistent with the constraint function boundary in Fig. 7.9a. Figure 7.9d depicts the contour plot of the CPI function, which is the product of the PI function and the PoF function. The maximum value of the CPI function is close to the true optimal feasible solution. By maximizing the CPI function to select new sample points, an optimal solution will be obtained that will approach the true optimal feasible solution.

7.1.2.4 Procedures for the SBO Algorithm Driven by the CPI Criterion

The procedures for the SBO algorithm driven by the CPI criterion are the same as those for the SBO algorithm driven by the CEI criterion, except that the search for new sample points is guided by maximizing the CPI criterion (as shown in Algorithm 7.4) instead of the CEI criterion.

Algorithm 7.4 Procedures for the SBO algorithm driven by the CPI criterion

Require: initial sample points $\mathbf{X}$, objective function $\mathbf{y}$ and constraint functions $\{\mathbf{g}_1, \mathbf{g}_2, \ldots, \mathbf{g}_c\}$

Find: the optimal solution $(\mathbf{x}_{\min}, y_{\min})$

1. **while** the stopping criterion is not satisfied **do**
   2. Construct a kriging model for the objective function based on the set of sample points $(\mathbf{X}, \mathbf{y})$
   3. **for** $i = 1$ to c **do**
   4. Construct the i-th kriging model for the i-th constraint function based on the set of sample points $(\mathbf{X}, \mathbf{g}_i)$ $(i = 1, 2, \ldots, c)$
   5. **end for**
   6. **if** no feasible solution exists among the current sample points **then**
   7. $x^{(u)} = \arg\max \ \mathrm{PoF}(\mathbf{x})$
   8. **else**

$x^{(u)} = \arg\max \ \mathrm{CPI}(\mathbf{x})$

end

Calculate the true response values of the objective function and each constraint function at the new sample point $\mathbf{x}^{(u)}$

$\mathbf{X} \leftarrow \mathbf{X} \cup \mathbf{x}^{(u)}$

$\mathbf{y} \leftarrow \mathbf{y} \cup y\left(\mathbf{x}^{(u)}\right)$

for $i = 1$ to c **do**

$\mathbf{g}_i \leftarrow \mathbf{g}_i \cup g_i\left(\mathbf{x}^{(u)}\right)$

end for

if no feasible solution exists among the current sample points **then**

Output "Cannot find any feasible solution!"

else

Update the current optimal solution to $(x_{\min}, y_{\min})$

end if

end while

The process of using the SBO algorithm driven by the CEI criterion to solve the constrained Branin problem (Forrester et al. 2008) is illustrated in Fig. 7.10. In the left column, the contours of the constrained Branin function and its boundary (represented by the blue lines in Fig. 7.10) are shown; in addition, the true optimal feasible solution is denoted by the blue stars, and the white dots indicate the sample points used to construct the kriging models. In the right column, the contour maps from the first three iterations are depicted, and the red dots indicate the maximum values of the CPI function, which determine the selection of the newly added sample points. In the first iteration, since the selected new sample point is in the feasible domain, the search tends towards the minimum region of the constrained Branin problem in the second iteration. From Fig. 7.10e, we can see that the second newly added sample point is very close to the true optimal feasible solution.

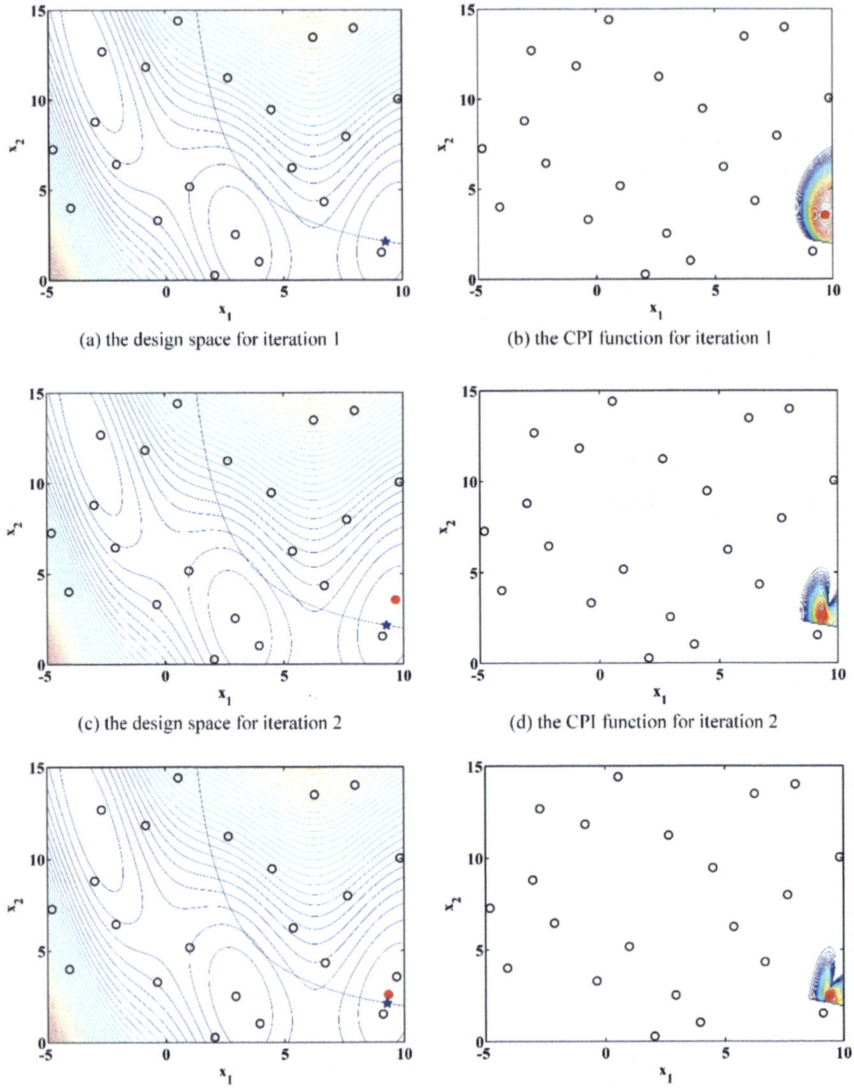

(a) the design space for iteration 1

(b) the CPI function for iteration 1

(c) the design space for iteration 2

(d) the CPI function for iteration 2

(e) the design space for iteration 3

(f) the CPI function for iteration 3

Fig. 7.10 The iterative process driven by the CPI criterion for the constrained Branin problem

7.1.3 *Lower Confidence Bound Criteria*

7.1.3.1 The Unconstrained Lower Confidence Bound (LCB) Criterion

The EI criterion can be used as the infill criterion to search for new sample points, as discussed in Sect. 7.2.1. When the predicted response value is smaller than the current minimum response value and the prediction error is small, the search will focus on the neighbourhood of the current minimum response value, emphasizing a local exploitation; when the predicted response value approaches the current minimum response value and the prediction error is large (that is, the accuracy of prediction at the unobserved point is low), the search will focus on regions with low prediction accuracy, emphasizing a global exploration. Therefore, when the prediction error at the unobserved point is large and/or the prediction response value at the unobserved point is smaller than the current minimum response value, the value of the EI function is large. Maximizing the EI function to obtain the next sample point can balance the relationship between the global and local search capabilities. However, to apply the EI criterion, a sufficient local search must be performed before the global search, resulting in a large amount of computation.

The LCB criterion is formulated as follows (Laurenceau et al. 2010):

$$\mathrm{LCB}(\mathbf{x}) = \hat{y}(\mathbf{x}) - b \cdot \hat{s}(\mathbf{x}) \tag{7.18}$$

where $\hat{y}(\mathbf{x})$ is the response value predicted by the kriging model at the unobserved point; and b is an equilibrium constant defined by the designers, which plays an important role in balancing the relationship between the global and local search capabilities. When $b = 0$, minimizing the LCB criterion is equivalent to minimizing $\hat{y}(\mathbf{x})$; consequently, the search is local in nature, focusing on the current minimum value of the objective function. When $b \to \infty$, the effect of $\hat{y}(\mathbf{x})$ is negligible, and minimizing the LCB criterion is equivalent to maximizing the prediction standard deviation $\hat{s}(\mathbf{x})$; consequently, the search is global in nature, focusing on areas with poor prediction accuracy. In addition, the problem of maximizing the LCB function can be converted into a minimization problem by multiplying by -1. The physical meaning of minimizing Eq. (7.18) can be seen as a combination of two objectives: minimizing $\hat{y}(\mathbf{x})$ while maximizing $\hat{s}(\mathbf{x})$. b is selected based on experience and can simply be defined as $b = 1$ to balance global exploration with local exploitation.

Figure 7.11 presents the LCB function (represented by the blue dotted line) for the one-dimensional function (Forrester et al. 2008). The true function is shown by the red line, and the kriging model built based on the four initial sample points (represented by the pink dots) is shown by the black dashed line. From Fig. 7.11,

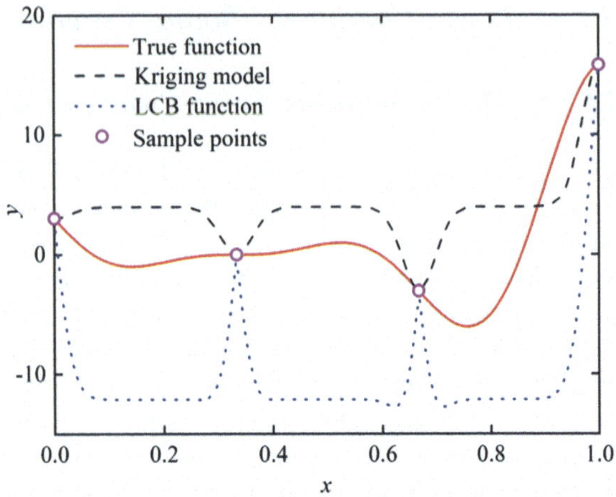

Fig. 7.11 The LCB function for the one-dimensional function

we can see that the value of the LCB function is always equal to or smaller than the value predicted by the kriging model. Since the prediction errors at sample points are 0, the LCB function at a sample point is equal to the prediction response. At unobserved points, the value of the LCB function is smaller than the prediction response because of the prediction error.

7.1.3.2 Procedures for the SBO Algorithm Driven by the LCB Criterion

The process of the SBO algorithm driven by the LCB criterion is similar to that of the SBO algorithm driven by the EI criterion, as described in Sect. 7.1.2.2. The procedures of SBO algorithm is summarized in Algorithm 7.5. First, the design variables and the design space A are identified. The initial kriging model is built based on initial sample points that are generated in the design space using a design of experiments method. The initial trust region T_1 is the entire design space A. Then, in each iteration of the loop, a new sample point is selected by minimizing the LCB criterion. After the true response value at the new sample point is calculated, this sample point is added to the sample set, and the kriging model is updated. In addition, the trust region is updated based on the currently known information. Once the stopping criterion is satisfied, the iterative process stops, and the optimization process ends. The optimal solution obtained in lines 7–8 is the optimal solution to the original problem. If the stopping criterion is not satisfied, the optimization algorithm will return to line 2.

During the optimization process, the LCB function is treated as the objective function. The solution obtained by minimizing the objective function is chosen as the next sample point to be added. Then, the kriging model is updated. This process is repeated until convergence is reached. When b is small, the new sample points are selected with a focus on the neighbourhood of the current minimum value of the objective function, and the overall search direction is biased towards the local region. When b is large, the nature of the search is continuously transformed between global and local, with an overall bias towards a global search, until convergence is reached.

Algorithm 7.5 Procedures for the SBO algorithm driven by the LCB criterion

---

Require: initial sample points $(\mathbf{X},\mathbf{y})$

Find: the optimal solution $(\mathbf{x}_{\min}, y_{\min})$

1. **while** the stopping criterion is not satisfied **do**
2. Construct a kriging model based on the set of sample points $(\mathbf{X},\mathbf{y})$
3. $x^{(u)} = \arg\min \ \mathrm{LCB}(\mathbf{x})$
4. Calculate the true response value $y\left(\mathbf{x}^{(u)}\right)$ at the new sample point $\mathbf{x}^{(u)}$
5. $\mathbf{X} \leftarrow \mathbf{X} \cup \mathbf{x}^{(u)}$
6. $\mathbf{y} \leftarrow \mathbf{y} \cup y\left(\mathbf{x}^{(u)}\right)$
7. $y_{\min} \leftarrow \min(y)$
8. $x_{\min} \leftarrow \mathbf{x} \in \mathbf{X}; y(\mathbf{x}) = y_{\min}$
9. **end while**

The optimization process driven by the LCB criterion is demonstrated here for the one-dimensional function (Forrester et al. 2008). The first three iterations and the last iteration are shown in Fig. 7.12. In the left column, the true function (red lines) and the kriging models (black dashed lines) built based on the sample points (white dots) are depicted. In the right column, the LCB function for each iteration is shown, and the calculated minimum values of the LCB functions are indicated by pink dots. The minimum value of the LCB function in Fig. 7.12 is obtained at the point indicated by the pink dot, and this point is added to the set of sample points and used to reconstruct the kriging model for the next iteration until the iterative process ends. Once nine new sample points have been added, the stopping criterion is satisfied, and the final optimal solution is obtained.

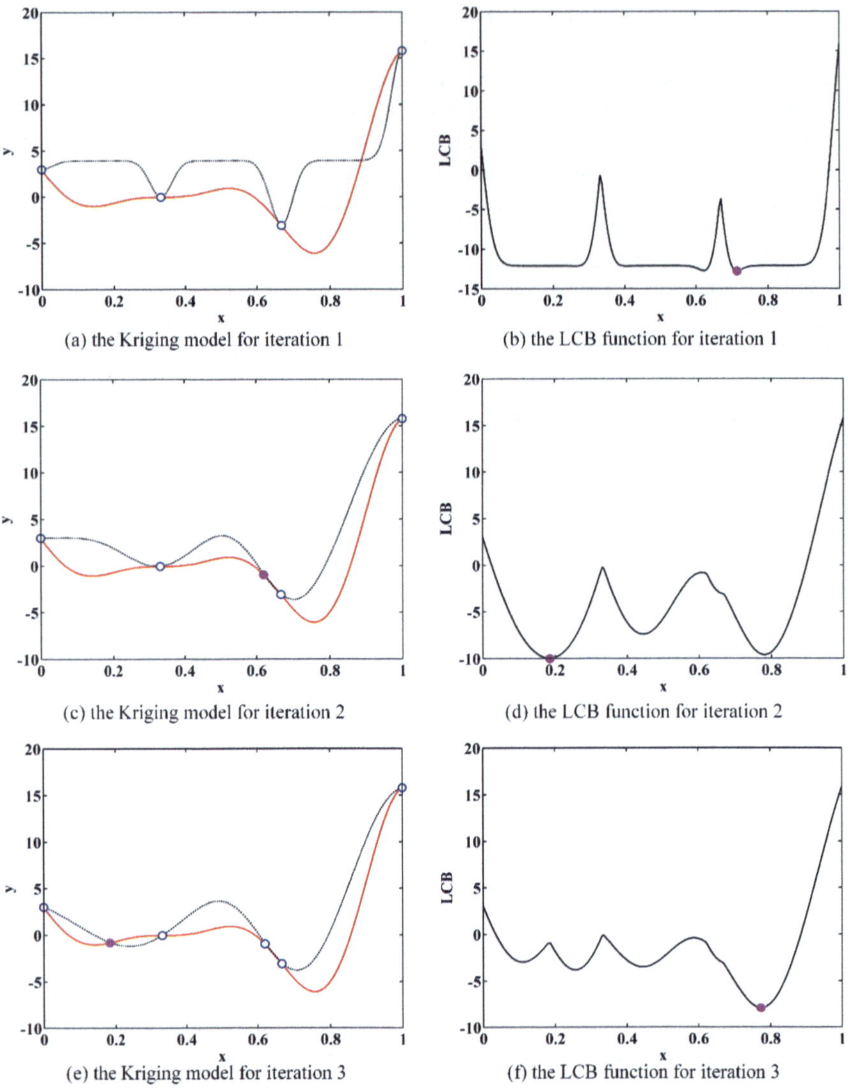

Fig. 7.12 The iterative process driven by LCB iteration on the one-dimensional function

7.1.3.3 The Constrained Lower Confidence Bound (CLCB) Criterion

The constrained lower confidence bound (CLCB) criterion has been proposed to solve constrained optimization problems. It can be regarded as a combination of the PoF criterion and the LCB criterion. There are some similarities between the CEI

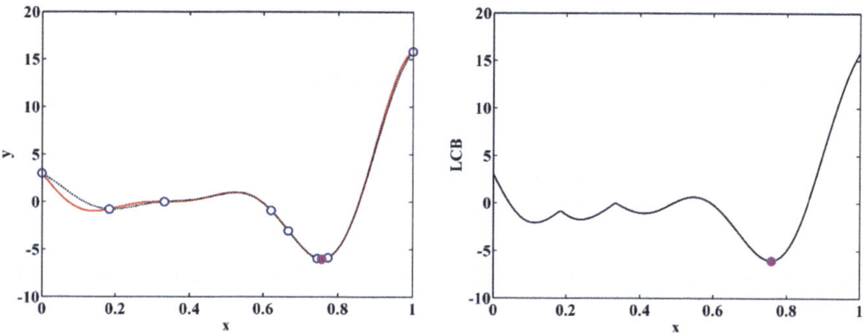

(g) the Kriging model for iteration 10 (h) the LCB function for iteration 10

Fig. 7.12 (continued)

and CLCB criteria, but there are also slight differences between them since the CLCB criterion considers different forms of the PoF function depending on the sign of the LCB function:

$$\mathrm{CLCB}(\mathbf{x}) = \begin{cases} \mathrm{LCB}(\mathbf{x}) \prod_{j=1}^{c} \frac{1}{\Pr(G_j(\mathbf{x}) \le 0)} & \text{if} \quad \mathrm{LCB}(\mathbf{x}) > 0 \\ \mathrm{LCB}(\mathbf{x}) \prod_{j=1}^{c} \Pr(G_j(\mathbf{x}) \le 0) & \text{if} \quad \mathrm{LCB}(\mathbf{x}) \le 0 \end{cases} \tag{7.19}$$

where $\mathrm{LCB}(\mathbf{x})$ is the LCB function without considering the constraint functions and $\Pr(G_j(\mathbf{x}) \le 0)$ is the probability that the point $\mathbf{x}$ satisfies the j-th constraint function. Thus, $\prod_{j=1}^{c} \Pr(G_j(\mathbf{x}) \le 0) = \prod_{j=1}^{c} \Phi\left(\frac{-\hat{g}_j(\mathbf{x})}{e_j(\mathbf{x})}\right)$, which is called the PoF function, represents the probability that the point $\mathbf{x}$ satisfies all constraint functions. Here, c represents the number of constraint functions.

The value of the PoF function is always greater than or equal to 0; if there is no feasible sample point, the values of the PoF function at all sample points are zero, while those at unobserved points are greater than zero. Therefore, the sign of the CLCB function is determined by the sign of the LCB function. If $\mathrm{LCB}(\mathbf{x}) > 0$, then $\mathrm{CLCB}(\mathbf{x}) > 0$; if $\mathrm{LCB}(\mathbf{x}) \le 0$, then $\mathrm{CLCB}(\mathbf{x}) \le 0$. To minimize the CLCB function, the search is performed by minimizing $\mathrm{LCB}(\mathbf{x})$, that is, minimizing $\hat{y}(\mathbf{x})$ and maximizing $\hat{s}(\mathbf{x})$. Thus, the search will tend towards areas with $\mathrm{LCB}(\mathbf{x}) \le 0$. In addition, the value of the PoF function will be greater when all constraint functions are satisfied than it will be when at least one constraint function is not satisfied. Thus, when minimizing the CLCB function, the search process will tend to select a point

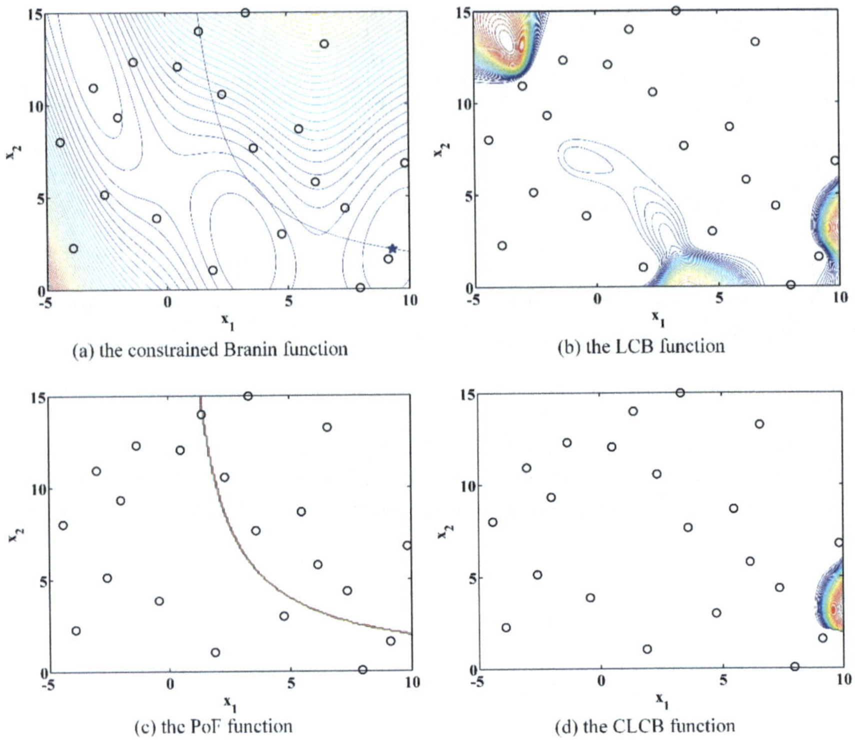

Fig. 7.13 The constrained Branin problem and its corresponding LCB, PoF and CLCB functions

where $\mathrm{LCB}(\mathbf{x}) \leq 0$ and all constraint functions are satisfied (when $\mathrm{LCB}(\mathbf{x}) \leq 0$, the greater value of the PoF function will make the value of the CLCB function smaller). In contrast, points at which $\mathrm{LCB}(\mathbf{x}) > 0$ and at least one constraint function is not satisfied (when $\mathrm{LCB}(\mathbf{x}) > 0$, the smaller value of the PoF function will increase the value of the CLCB function) can more easily be eliminated.

The constrained Branin problem is used as an example here to explain the basic idea of the CLCB criterion. In Fig. 7.13, the contour maps of the constrained Branin problem and its LCB, PoF and CLCB functions are given. The blue solid line in Fig. 7.13a is the constraint boundary of the constrained Branin problem; the area on the upper right side of the boundary is the feasible region, while the area on the lower left side is the infeasible region. The sample points used to build the kriging model are indicated by white dots. The true optimal feasible solution to the constrained Branin problem is represented by a blue star. The more darkly an area is coloured, the larger the value of the function in that area. The three maximum regions of the LCB function in Fig. 7.13b are quite consistent with three minimum regions of the constrained Branin problem. The PoF function in Fig. 7.13c is similar to the boundary defined by the constraint function of the constrained Branin problem in Fig. 7.13a. In Fig. 7.13d, the maximum region of the CLCB function is exactly coincident with the location of the true optimal solution. The CLCB criterion is the product of the LCB function and the PoF function. By minimizing the CLCB function, the search process can be easily and quickly driven towards the neighbourhood of the true optimal feasible solution.

7.1.3.4 Procedures for the SBO Algorithm Driven by the CLCB Criterion

When the CLCB criterion is used to select the next sample point, the sign of the LCB function determines the form of the PoF function that is used when minimizing the CLCB criterion. Therefore, the procedures for the SBO algorithm driven by the CLCB criterion are different from those for the SBO algorithm driven by the CEI criterion. Again, there are two main steps in the optimization process: first, generating the initial sample set using a design of experiments method, and then, selecting a new sample point in each iteration of the loop by minimizing the CLCB criterion. The iterative process does not terminate until the stopping criterion is met.

The optimization process of the constrained SBO algorithm driven by the CLCB criterion is shown in Algorithm 7.6. First, the true response values of both the objective function and the constraint functions at the initial sample points are calculated through simulations and analysis. Therefore, in lines 2–4, kriging models are built not only for the objective function but also for the constraint functions. Then, a new sample point is selected by minimizing the CLCB function. Next, the true response values of the objective function and the constraint functions at the selected sample point are calculated. After adding this point to the set of existing sample points and before selecting the next new sample point, it is necessary to judge whether there is a feasible solution in the current set of sample points if the stopping criterion has not yet been met. In the next iteration, the kriging models are reconstructed based on the new sample set. When the iterative process ends, the current optimal solution is adopted as the final optimal solution and is output.

Algorithm 7.6 Procedures for the SBO algorithm driven by the CLCB criterion

Require: initial sample points $\mathbf{X}$, objective function y and constraint functions $\{\mathbf{g}_1, \mathbf{g}_2, \ldots, \mathbf{g}_c\}$

Find: the optimal solution $(\mathbf{x}_{\min}, y_{\min})$

1. **while** the stopping criterion is not satisfied **do**
2. Construct a kriging model for the objective function based on the set of sample points $(\mathbf{X}, \mathbf{y})$
3. **for** $i = 1$ to c **do**
4. Construct the i-th kriging model for the i-th constraint function based on the set of sample points $(\mathbf{X}, \mathbf{g}_i)$ $(i = 1, 2, \ldots, c)$
5. **end for**
6. $x^{(u)} = \arg\min \ \mathrm{CLCB}(\mathbf{x})$
7. Calculate the true response values of the objective function and each constraint function at the new sample point $\mathbf{x}^{(u)}$
8. $\mathbf{X} \leftarrow \mathbf{X} \cup \mathbf{x}^{(u)}$
9. $\mathbf{y} \leftarrow \mathbf{y} \cup y(\mathbf{x}^{(u)})$
10. **for** $i = 1$ to c **do**
11. $\mathbf{g}_i \leftarrow \mathbf{g}_i \cup g_i(\mathbf{x}^{(u)})$
12. **end for**
13. Update the current optimal solution to $(x_{\min}, y_{\min})$
14. **end while**

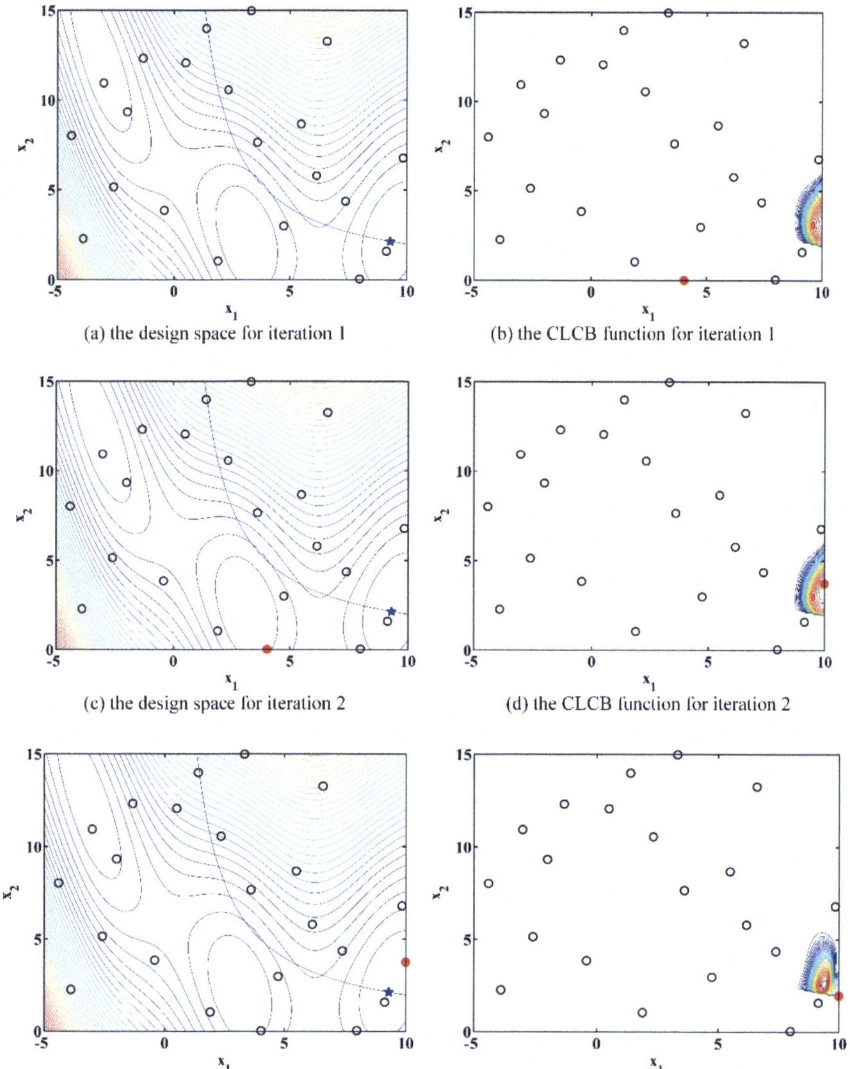

(a) the design space for iteration 1
(b) the CLCB function for iteration 1
(c) the design space for iteration 2
(d) the CLCB function for iteration 2
(e) the design space for iteration 3
(f) the CLCB function for iteration 3

Fig. 7.14 The iterative process driven by CLCB iteration on the constrained Branin function

To demonstrate the iterative process of the SBO algorithm driven by the CLCB criterion, the constrained Branin problem is used as an example. In the first iteration, the initial kriging model is built based on the initial sample points (white dots in Fig. 7.14a). Additionally, the CLCB function is obtained, making full use of the information on the predicted values and prediction errors. Then, the minimum value of the CLCB function is found (represented by the red dot in Fig. 7.14b). Since the stopping criterion has not been satisfied, the algorithm proceeds to the second

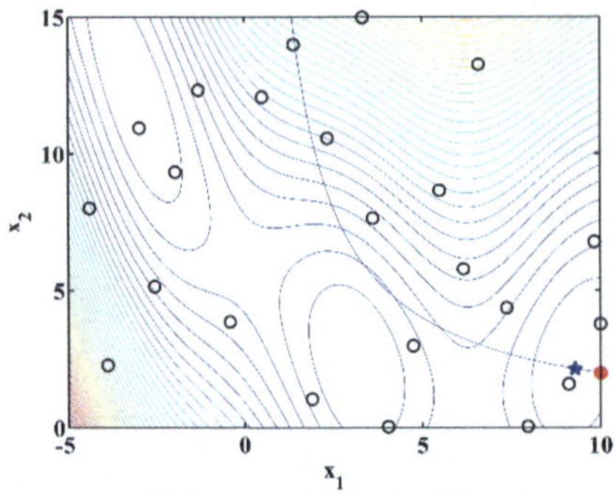

(g) the design space for iteration 4

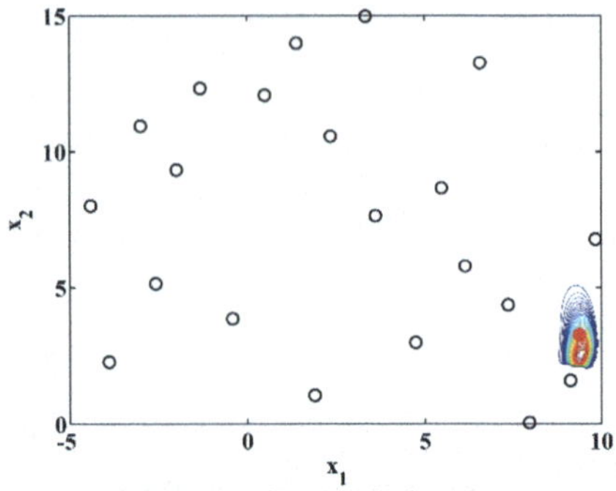

(h) the CLCB function for iteration 4

Fig. 7.14 (continued)

iteration. The newly selected point (represented by the red dot in Fig. 7.14c) is added to the existing sample points, representing the first update to the sample set. Note that the solid blue line in Fig. 7.14c represents the constraint boundary of the constrained Branin problem. The region on the left side of the boundary is the infeasible domain, and the region on the right side is the feasible domain. Thus, the first newly added sample point is infeasible. Therefore, the most important task in the second iteration is to search for a new sample point in the feasible region. Consequently, the PoF function is the dominant driver of the search process, based on which the second new sample point is obtained (represented by the red dot in Fig. 7.14d). It is obvious that this new point is in the feasible domain. Therefore, the search process in the third iteration is dominated by the LCB function, and the third new sample point (represented by the red dot in Fig. 7.14f) is selected from the minimum feasible domain. When the algorithm proceeds to the fourth iteration, the third newly added sample point selected based on the CLCB criterion is very close to the true optimal feasible solution, as shown in Fig. 7.14g.

7.2 Surrogate-Model-Based Robust Design Optimization

In deterministic optimization problems, the design variables, objective functions and constraints are deterministic and without multi-source uncertainties arising from engineering design and the product manufacturing process (Eldred et al. 2002). Consequently, the optimal solution may be very sensitive to these uncertainties (Chen et al. 2016). Robust optimization (RO) methods have been proposed to obtain solutions that are both optimal and relatively insensitive to input uncertainty. RO methods can be classified into two types: probabilistic approaches and deterministic approaches. In probabilistic approaches, RO is performed based on the probability distributions of the variable variations, usually the means and variances of the uncertain variables (Du and Chen 2004; Du et al. 2008; Chen et al. 2014; Lim et al.

2014; Li et al. 2016). Deterministic approaches, on the other hand, incorporate non-statistical indexes such as gradient information (Taguchi 1978; Renaud 1997; Lee and Park 2001; Kim et al. 2010; Papadimitriou and Giannakoglou 2013) or sensitivity region information (Gunawan and Azarm 2004, 2005a, b; Li et al. 2006, 2009b, 2010; Hu et al. 2011; Zhou et al. 2012, 2015c; Zhou and Li 2014; Cheng et al. 2015c) into the original optimization problem to obtain a robust optimum. With the rapid development of computer capabilities and speed, computational simulation methods, e.g. computational fluid dynamics and finite element analysis, become widely used in RO methods in place of computationally intensive real-life experiments for performance evaluation. However, directly using computer simulations along with an optimizer to evaluate many design alternatives when exploring the design space in search of a robust optimum can be computationally prohibitive (Zhang et al. 2012). Therefore, surrogate-model-based RO approaches have recently attracted significant attention. In this chapter, the typical surrogate-model-based probabilistic and deterministic RO approaches are introduced.

7.2.1 Surrogate-Model-Based Probabilistic Robust Optimization (RO) Approaches

7.2.1.1 Basic Concept of Probabilistic RO

When the design variables x and parameters (noise factors) z of a robust design problem are random variables (Zhou et al. 2015b), it is very effective to adopt a stochastic model and an optimization method for random variables to solve the robust design or RO problem. The optimal design of random engineering variables (Zhu et al. 2009) is a branch of optimization technology that has been developed in recent decades. It provides an important set of tools for applying optimization technology, probability theory, mathematical statistics and computer technology to engineering problems to cope with various random factors. When an engineering problem involves random design variables and random parameters, the response value $\mathrm{y} = y(x, z)$ is also a random variable (Coello 2000). In this case, the general form of the stochastic optimization design model is as follows:

$$\begin{aligned} & x \in (\Omega, \mathcal{T}, P) \subset R^n \\ & opt. \quad F_0(x) = F\{y(x, z)\} \\ & s.t. \quad F_j(x) = G\{g_j(x, z)\} \leq 0, \mathrm{j} = 1, 2, \ldots, m \\ & z \in (\Omega, \mathcal{T}, P) \subset R^n \end{aligned} \tag{7.20}$$

where x is the n-dimensional design variable vector; z is the k-dimensional random parameter vector; $F\{\cdot\}$ and $G\{\cdot\}$ are generic functions with some probabilistic meaning, for example, the expectation value, variance, or probability; and $(\Omega, \mathcal{T}, P)$ is the probability space. This means that both x and z belong to the probability space.

A collection of design variables x is considered to be the optimal solution x^* to the model if it belongs to the set

$$\mathcal{D} = \{x | F_j(x) \leq 0, \mathrm{j} = 1, 2, \ldots, \mathrm{m}\} \tag{7.21}$$

Its probability distribution and distribution parameters should be known, and x should satisfy the condition

$$F_0(\bar{x}^*) = minF_0(x) \tag{7.22}$$

There are two basic ways to solve the model introduced above. The first way is to transform the stochastic model into a deterministic one, which is feasible for cases in which the random variables and stochastic functions are normally distributed and the dispersion coefficients of the variables are small (Han et al. 2012). The second way is to use a stochastic method to solve the random model directly. Methods of this kind are not limited to any particular forms of the variable distributions or of the correlations between the random variables (Zhou et al. 2017). To solve problems of engineering programme design, several basic concepts and definitions related to stochastic problems need to be introduced, based on which the challenges related to modelling and solution methods for robust design can be reasonably addressed (Oberkampf et al. 2001). Therefore, in the first few parts of this section, several basic problems related to stochastic modelling are introduced. It is worth mentioning that when a surrogate model is applied, the uncertainty of this model, as discussed in Chap. 5, also needs to be considered.

The variations encountered in practical engineering problems can be divided into two general types, as follows. Variations of the first type are related to the uncertainty of the data obtained by measuring or testing physical or mechanical processes. Variations of the second type are caused by random processes or factors of chance. These types of randomness can be represented in terms of random variables that obey a particular probability distribution (Gano et al. 2006a).

(1) Random analysis of the design variables

During the design process, we need to adjust the mean value and tolerance of each design variable to reduce the deviation of the quality index y and the fluctuation value ε_y (Haftka 1991). The design variables can be expressed as

$$x^T = (x_1, x_2, \ldots, x_n) = (\Omega, \mathcal{T}, P) \subset R^n \tag{7.23}$$

where Ω is a random event, $\mathcal{T}$ represents the whole random event set, and P represents the probability. In this way, the design variables can be expressed as

$$x_i = \bar{x}_i + t_i, i = 1, 2, \ldots, n \tag{7.24}$$

Considering the stochastic independence of the design variables, it is assumed that t_i follows the normal distribution $N(0, \sigma)$; therefore, its mean, variance and covariance are

$$E(t_i) = 0; \mathrm{Var}(t_i) = \sigma_{x_i}^2 = (\delta_i \bar{x}_i)^2; \mathrm{Cov}(t_i, t_i) = 0, j \neq i \tag{7.25}$$

where δ_i is a deviation function. The tolerance range of design variable x_i is assumed to be $\Delta\bar{x}_i = |\Delta x_i^+| = |\Delta x_i^-|$ $(i = 1, 2, \ldots, n)$. Based on the three-sigma quality principle, the relationship between the tolerance Δx_i and the standard deviation σ_{x_i} of this design variable is presented as follows:

$$\sigma_{x_i} = \frac{\bar{x}_i + \Delta x_i - (\bar{x}_i - \Delta x_i)}{6} = \frac{\Delta x_i}{3} \tag{7.26}$$

Since $|t_i| \leq \Delta x_i$, in the general case, the tolerance for design variable x_i should be specified; that is, Δx_i should also be treated as a design variable during the design process. In rare cases, a deviation function δ_i could also be set as a design variable. A more intuitive illustration is provided in Fig. 7.15.

Theoretically, the distribution types and the design variables should be adjusted simultaneously when optimally solving the stochastic model. However, it is impossible to achieve precise results (Gu et al. 2000). As a result, depending on the characteristics of the mechanical product design process, the designers could refer to statistical test experience and sample test experience for critical components, data on the parameters of similar components, or purely subjective inference (usually, the first option chosen will be a normal distribution) to determine the types of distributions that the design variables obey. Then, by adjusting the distribution parameters (shape, size and position parameters) or characteristic values (mean and variance), the optimal solution to the problem can be gradually acquired (Kerschen et al. 2006). In particular, when the latter approach is applied, a random design variable can be represented in terms of its mean and deviation, which are consistent with the nominal values used in mechanical design (Madsen et al. 2006). In this way, when the deviation coefficient of some design variable is known, its mean value can be used to assist in the iterative computations for optimization.

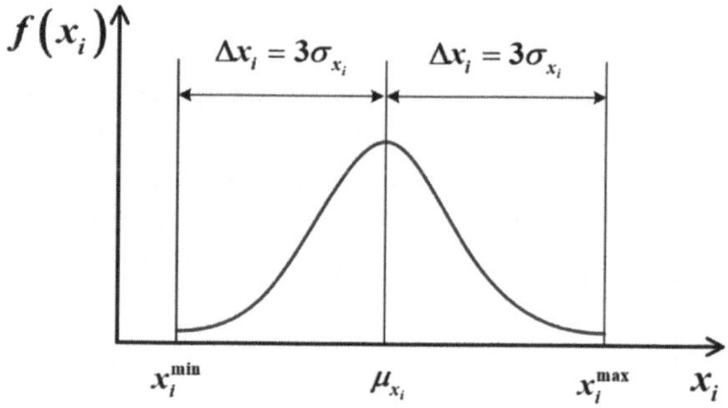

Fig. 7.15 Normally distributed design variables

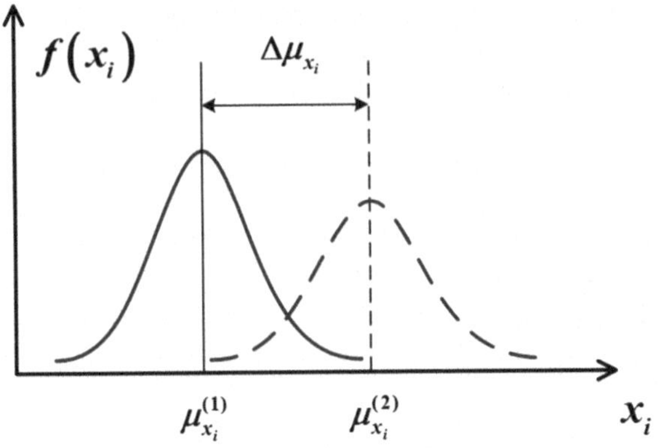

Fig. 7.16 Iteration with the mean values of random design variables

$$\Delta\bar{x}_i^{(k+1)} = \bar{x}_i^{(k)} + \alpha_i S_i^{(k)},\ i = 1, 2, \ldots, n \tag{7.27}$$

In this way, the process can be considerably simplified, as illustrated in Fig. 7.16.

From the above, it is clear that the design variable x_i can be regarded as representing a random variation within the tolerance range $\left[\Delta x_i^-, \Delta x_i^+\right]$ for dealing with the concept of 'randomness' (Arendt et al. 2013). For example, each geometric dimension of a mechanical part will obey a normal distribution with mean μ and variance σ^2; the nominal value of each dimension can be set to the corresponding mean value μ, and the tolerance range can be obtained by calculating $\pm k\sigma$, where $k = 1, 2, \ldots$. For instance, when $k = 2$, the true value will be within the range of $[\mu - 2\sigma, \mu + 2\sigma]$ with 95.45% probability. When $k = 3$, the probability reaches 99.73%. When the tolerance is unknown, the optimal solution to the problem can also be found by adjusting the mean $\overline{x_i}$ of the random design variable and controlling the deviation coefficient δ_i to represent the tolerance ($\Delta x_i = 3\mu_x \delta_i$). This approach is beneficial for studying problems such as maximum tolerance design, the optimal allocation of tolerance and the most economical tolerance.

When a design variable has a definite value, it can be regarded as a special type of random variable with a dispersion coefficient of zero.

(2) Random analysis of the uncertainty parameters

Uncertainty (randomness or fuzziness) is presented in both internal and external noise factors (Zhou et al. 2016b). Examples include the randomness of working loads, the physical and mechanical properties of materials, wind forces and rainfall as well as the fuzziness of product quality evaluations, the intensity grading of phenomena such as wind strength and earthquakes, etc. (Apley et al. 2006). Noise factors whose distribution types and parameters are already known during the design process are called random parameters z, which can be represented as

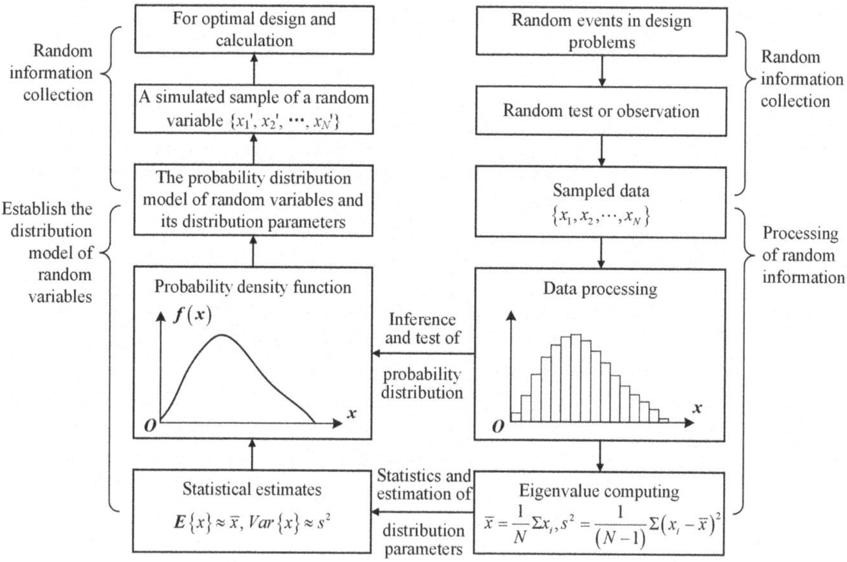

Fig. 7.17 Collection, processing, simulation and application process for random information

$$z^T = (z_1, z_2, \ldots, z_k) \in (\Omega, \mathcal{T}, P) \subset R^k \tag{7.28}$$

where every random parameter z_i is a function of certain distribution parameters α_i, β_i and γ_i (Gano et al. 2006b):

$$z_i = \phi(\alpha_i, \beta_i, \gamma_i), i = 1, 2, \ldots, k \tag{7.29}$$

Here, $\phi(\cdot)$ is the distribution function, and α_i, β_i and γ_i represent its position, shape and size parameters, respectively. The k random parameters are generally mutually independent, but in some cases, their cross-correlations cannot be ignored.

For practical problems, it is generally most convenient to determine the probability density function of random parameter z_i because this function directly reflects the parameter's probabilistic statistical properties, based on which a simulated sample of the corresponding random variable can be generated (Zhu et al. 2015). Figure 7.17 summarizes the process of collecting, processing, simulating and applying such random information.

In a practical case, if only a few trials or observations of random events are conducted, say fewer than ten, then the analogy method and the feature point method can be used to approximate the probability density function (Bahrami et al. 2016). The probability density function can also be chosen based on experience. Examples of several typical theoretical distributions are given in Table 7.1.

In design optimization, since probability density functions are always obtained in an analytical, it will not be challenging to generate random samples from these

Table 7.1 Application examples of several theoretical distributions

Theoretical probability distribution	Sampling application
Triangular distribution	Part size deviation, roughness
Normal distribution	Measurement error, manufacturing dimensional deviation, rigidity, material strength limit, elasticity modulus, systematic error, random error, fracture toughness, metal wear, applied loading, air humidity, expansion coefficient, clearance error
Lognormal distribution	Strength limit of alloy material, the fatigue life of materials, rainfall intensity, completion time, spring fatigue strength, corrosion, corrosion coefficient, the inner pressure of container, metal cutting tool durability, system trouble-free working time, tooth bending strength and contact fatigue strength
Weibull distribution	Fatigue strength, fatigue life and wear life in mechanical engineering, radial runout of a shaft, system life
Rayleigh distribution	A special occasion of Weibull distribution. Tolerance of shape and position (such as conicity, rectangularity, flatness, ovality, eccentricity)
Extreme value distribution	Various types of load, the extremum value of load (the maximum or minimum)
Beta distribution	A distribution suitable for some bounded random variable ($a<x<b$). The probability density function curve shapes differ in the value of α_1, α_2
Binomial distribution	Typhoon distribution, river pollution distribution, yield distribution
Poisson distribution	The distribution of hard stones in the soil, statistical quality inspection, public service, failure rate

functions(Zhou et al. 2016a). In the case in which the probability distribution is known, the simplest way to generate a random parameter sample is through redundancy. The steps are as follows:

Step 1: Obtain the probability density function based on a small number of random samples obtained through experiment or observation.
Step 2: Determine whether the probability density function has boundaries. If it does not, the truncation limits at the two ends can be obtained in accordance with the specified precision requirement.
Step 3: Calculate the maximum $f_{\max}$ of the probability density function $f(z)$.
Step 4: Calculate the upper and lower bounds on the value of the random parameter z, denoted by $z_{\max}$ and $z_{\min}$, respectively.
Step 5: Generate two random numbers, denoted by r_1 and r_2, from a uniform distribution on the interval [0, 1].
Step 6: Calculate the sample value

$$z = r_1(z_{\max} - z_{\min}) + z_{\min} \tag{7.30}$$

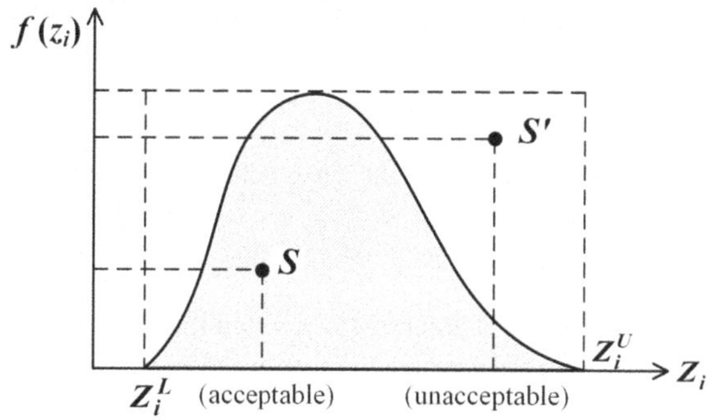

Fig. 7.18 Schematic diagram of the acceptance–rejection sampling method

and determine whether the following condition holds:

$$\mathrm{r}_2 \leq \frac{f(z)}{f_{\max}} \tag{7.31}$$

If Eq. (7.31) is true, then the sample z is accepted. Otherwise, the sample is rejected, and the process returns to Step 5. As shown in Fig. 7.18, the principle of this acceptance–rejection sampling method is very simple. It can be used for sampling one-dimensional random parameters as well as for random parameter vectors with or without correlations (Zhu et al. 2014). With the help of the acceptance–rejection sampling method, the sampling information required for calculating the random parameters of a random model can be obtained (Steuben and Turner 2015).

(3) Objective functions and constraints subject to uncertainty

In design optimization problems, when certain quality characteristics or technical indicators are functions of random design variables x and random parameters z, such a quality characteristic is called a stochastic design function (Simpson et al. 2001). It can be either a linear or a nonlinear function of one or more random variables. The general expression is as follows:

$$y = y(x, z) = \mathrm{y}(x_1, x_2, \ldots, x_n; z_1, z_2, \ldots, z_k) \in (\Omega, \mathcal{T}, P) \tag{7.32}$$

Since the objective functions and constraints in such models are usually established based on stochastic design functions, an objective function or constraint of this kind is also random (Tan 2015a). There are three commonly used metrics for measuring the randomness of such an objective function or constraint:

The expectation value (mean):

$$E\{y\} = E\{y(x, z)\} = E\{y(x_1, x_2, \ldots, x_n; z_1, z_2, \ldots, z_k)\} \tag{7.33}$$

The variance of the design function:

$$Var\{y\} = Var\{y(x,z)\} = Var\{y(x_1, x_2, \ldots, x_n; z_1, z_2, \ldots, z_k)\} \tag{7.34}$$

The probability that the design function satisfies the target value y_0:

$$P\{y \geq y_0\} = P\{y(x,z) \geq y_0\} \tag{7.35}$$

For a quality characteristic y, when the statistical average of $y - y_0$ is less than or equal to zero (or greater than or equal to zero), this is represented as

$$E\{y \geq y_0\} \leq 0 \tag{7.36}$$

A constraint expressed in this form is a mean-type constraint, where the predetermined target value y_0 may be either deterministic or random. We adopt the notation $g = y - y_0$; then,

$$E\{g\} = \int_{-\infty}^{+\infty} gf(g)\,dg = \int_{-\infty}^{+\infty} yf(y)dy - \int_{-\infty}^{+\infty} y_0 f(y_0)\,dy_0 \leq 0 \tag{7.37}$$

where $f(\cdot)$ is the function describing the probability density distribution of the random variable in the parentheses. Figure 7.19 presents a schematic illustration of a mean-type constraint.

Mean-type equality constraints are applicable only in certain special cases, such as when the average value of a design feature is required to meet specified technical requirements (Gluzman and Yukalov 2006). In the case of spring deflection, for example, the wire diameter, spring diameter, winding number and material properties are all random variables, meaning that the deflection is also random (Chen et al. 2005). However, the designer can specify that the overall average deflection of the spring should meet certain technical requirements.

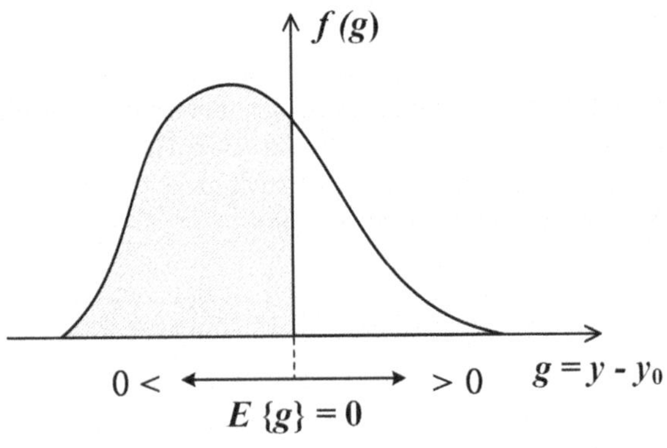

Fig. 7.19 Schematic illustration of a mean-type equality constraint

In the general case, we might obtain a transcendental equation of the following form as a stochastic equality constraint function:

$$h(x_1, x_2, \ldots, x_n; z_1, z_2, \ldots, z_k) = 0 \quad (7.38)$$

In this case, not only is it not advisable for the mean value to satisfy the equality constraint, but it is also impossible for each individual design variable to meet the constraint. Therefore, when building a stochastic design optimization model, we should not treat an expression of this kind as an equality constraint but rather should treat one of the design variables as a function of other random design variables and random parameters (Cheng et al. 2015c).

For a random design feature, the designer may specify that the probability of this feature satisfying $y - y_0 \leq 0\,(or \geq 0)$ should be greater than or equal to a certain preselected probability value α_0, represented as

$$P\{y - y_0 \leq 0\} \geq \alpha_0 \in [0, 1] \quad (7.39)$$

Such a constraint is called a probability constraint.

As in the case of a mean constraint, the target parameter y_0 of a probability constraint may be either deterministic or random. Mathematically, the probability value for such a constraint can be calculated as follows (Ghisu et al. 2011). Figure 7.20 presents the calculation diagram for a probability constraint.

Let $g = y - y_0$, and let $f(\cdot)$ represent the function describing the probability density distribution of the random variable in the parentheses. Then,

$$P\{g \leq 0\} = P\{y - y_0 \leq 0\} = \int_{-\infty}^{+\infty} f(y) \int_{y}^{+\infty} dy_0\, dy = \int_{-\infty}^{+\infty} f(y_0) \int_{-\infty}^{y_0} f(y)\, dy dy_0 \quad (7.40)$$

The concept of a probability constraint is geometrically illustrated in Fig. 7.21.

Fig. 7.20 Calculation diagram for a probability constraint

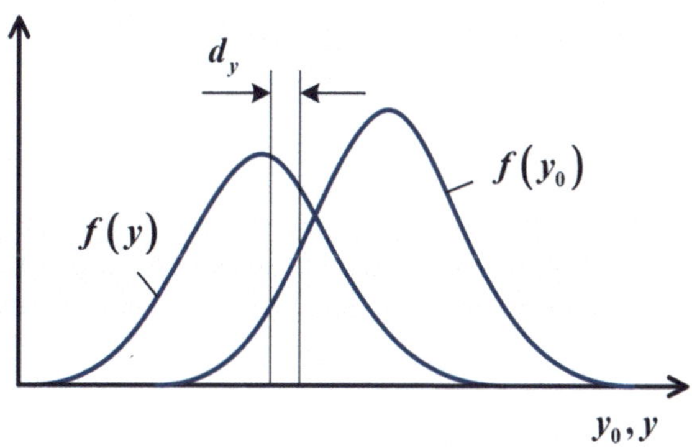

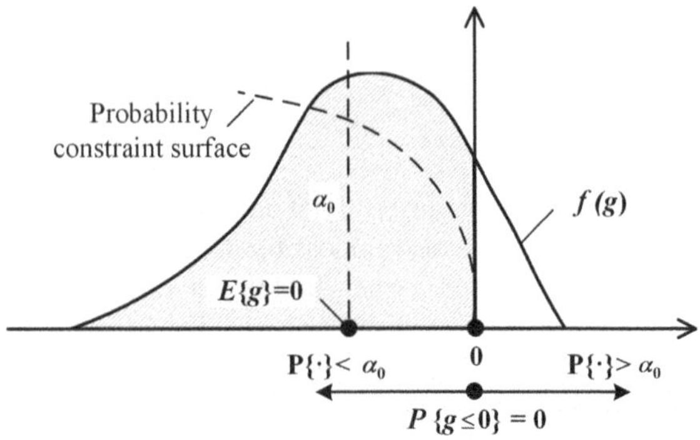

Fig. 7.21 Geometric relation between a probability constraint and a mean constraint

7.2.1.2 Probabilistic RO Considering the Uncertainties of the Design Variables and a Multi-fidelity Surrogate Model

(1) The uncertainty quantification process

A general mathematical model for a constrained RO problem can be expressed as follows:

$$\begin{aligned}
&Minimize : F(X) = \mu(f(X)) + c\sigma(f(X)) \\
&Subject\,to : G_i(X) = \mu(g_i(X)) + c\sigma(g_i(X)) \le 0 \\
&i = 1, 2, \ldots, L
\end{aligned} \tag{7.41}$$

where $X = [X_1, X_2, \ldots, X_q]$ is the random design variable vector, with q dimensions; $\mu(f(X))$ and $\sigma(f(X))$ are the mean and variance, respectively, of the objective function; $\mu(g_i(X))$ and $\sigma(g_i(X))$ are the mean and variance, respectively, of the *i*-th constraint; L is the number of constraints; and c is a constant value that reflects the risk attitude of the design. A larger c implies a more conservative attitude towards uncertainties. When $f(X)$ and $g(X)$ follow Gaussian processes, different choices of c lead to different confidence levels of the prediction intervals (Tan 2015b). For example, $c = 3$ represents a probability of 0.9987.

Here, the random design variable vector X can be split into a deterministic portion and an uncertain portion, as follows:

$$X = x + W \tag{7.42}$$

where $x = [x_1, x_2, \ldots, x_q]$ represents the deterministic variables and $W = [W_1, W_2, \ldots, W_q]$ represents the uncertainty of the design variables, which reflects the randomness of X. Here, it is assumed that W follows a Gaussian process with zero mean and a given covariance, i.e. $W \sim N(0, \sigma_x^2)$; therefore, X still follows a multivariate normal distribution $X \sim N(0, \sigma_x^2)$.

To solve the robust problem presented in Eq. (7.41), the means and variances of the objective and constraints must be evaluated (Younis and Dong 2010). In the following, expressions are presented for the mean and variance of the objective function under different circumstances; analogous expressions can be used for the constraints.

When a surrogate model is not applied, which means that the true response values of the objective function can be obtained, the mean and variance can be calculated as follows:

$$\mu_1(f(X)) = E[y_h(X)] = \int_w y_h(x+w)p(w)dw \tag{7.43}$$

$$\begin{aligned}\sigma_1^2(f(X)) &= Var[y_h(X)] = E\left[y_h^2(X)\right] - E[y_h(X)]^2 \\ &= \int_w y_h^2(x+w)p(w)dw - \left[\int_w y_h(x+w)p(w)dw\right]^2\end{aligned} \tag{7.44}$$

where $y_h(X)$ is the true response value and $p(w)$ is the joint distribution of w. When the random parameters w_i are independent and the marginal probability density functions p_{w_i} are given, the joint probability density function $p(w)$ can be calculated as follows:

$$p(w) = \prod_{i=1}^{d} p_{w_i}(w_i) \tag{7.45}$$

When a multi-fidelity (MF) surrogate model is constructed to be used in place of the high-fidelity (HF) model for obtaining the response values of the objective function, the mean and variance can be calculated (without considering the surrogate model) as follows:

$$\begin{aligned}\mu_2(f(X)) &= E\left[y_{mf}(X)\right] \\ &= E\left[y_{mf}(x+w)\right] \\ &= \int_w y_{mf}(x+w)p(w)dw\end{aligned} \tag{7.46}$$

$$\begin{aligned}\sigma_2(f(X)) &= Var\left(y_{mf}(X)\right) \\ &= E\left[y_{mf}^2(X)\right] - E\left[y_{mf}(X)\right]^2 \\ &= \int_w y_{mf}^2(x+w)p(w)dw - \left[\int_w y_{mf}(x+w)p(w)dw\right]^2\end{aligned} \tag{7.47}$$

where $y_{mf}(X)$ is the response value obtained from the MF surrogate model.

Using Eqs. (7.46) and (7.47) instead of Eqs. (7.43) and (7.44) to evaluate the mean and variance can significantly ease the computational burden. However, the MF surrogate model also introduces interpolation uncertainty as a result of the limited number of simulation runs, as illustrated in Chap. 5. This uncertainty is a type of model uncertainty, which is an epistemic uncertainty (Arendt et al. 2013; Hu and Mahadevan 2017). Treating the MF surrogate model as an accurate representation of the HF model and ignoring the additional prediction uncertainty of the MF model may result in non-optimal robustness (Cheng et al. 2015b). Hence, it is essential to quantify the combined effect of the uncertainties in the design variables and the MF surrogate model when evaluating the mean and variance.

For simplicity, $Y(X, V)$ is used to denote the response obtained from the MF surrogate model, where X represents the uncertainty of the design variables and V represents the uncertainty of the MF surrogate model itself (Zimmermann and Han 2010). Using statistical methods, the uncertainties of the design variables and the MF surrogate model are treated equally; then, the mean and variance of the objective function can be expressed as

$$\begin{aligned}\mu_2(f(X)) &= E[Y(X,V)] \\ &= \int_w y_{mf}(x+w)p(w)dw\end{aligned} \tag{7.48}$$

$$\begin{aligned}\sigma_3(f(X)) &= Var(Y(X,V)) \\ &= \int_w s^2\left(y_{mf}(x+w)\right)p(w)dw + Var\left[y_{mf}(x+w)\right]\end{aligned} \tag{7.49}$$

where $Var\left[y_{mf}(x+w)\right]$ is equal to $\sigma_2^2 f(X)$, which reflects the effects of the uncertainty of the design variables, and $s^2\left(y_{mf}(x+w)\right)$ is the mean square error (MSE) of the hierarchical kriging (HK) prediction, which reflects the effects of the uncertainty of the MF surrogate model. Thus, $\sigma_3^2(f(X))$ reflects the combined effects of the uncertainties of the design variables and the MF surrogate model (Xia et al. 2016). Since a surrogate model is used, $y_{mf}(x+w)$ can be inexpensively obtained. The Monte Carlo integration method can be used to calculate Eqs. (7.48) and (7.49). The concrete derivations of Eqs. (7.48) and (7.49) are presented below.

For simplicity, $Y(X, V)$ is used to denote the responses considering the uncertainty of the MF surrogate model, where X represents the uncertainty of the design variables and V represents the uncertainty of the MF surrogate model itself. The mean value of the objective function can be expressed as

$$\begin{aligned}
\mu_3(f(X)) &= E[y(X, V)] \\
&= E[E[Y(X, V)/V]] \\
&= E[E[Y(x+w, V)|V]] \\
&= E\left[\int_w Y(x+w, V)p(w)dw|V\right] \\
&= \int_w E[Y(x+w, V)|V]p(w)dw
\end{aligned} \tag{7.50}$$

Note that the integral $\int_w Y(x+w, V)p(w)dw|V$ in the fourth line exhibits randomness because of its functional dependence on V and not because of any dependence on w. The random effects of w are integrated out in this function. Because $E[Y(x+w, V)|V] = y_{mf}(x+w)$, one can obtain the mean value of the objective function as follows:

$$\begin{aligned}
\mu_3(f(X)) &= E[y(X, V)] \\
&= \int_w y_{mf}(x+w)p(w)dw
\end{aligned} \tag{7.51}$$

In the same manner, the variance of the objective function can be expressed as

$$\begin{aligned}
\sigma_3^2(f(X)) &= Var[Y(X, V)] \\
&= Var[Y(x+w, V)] \\
&= E\left[Y^2(x+w, V)\right] - E[Y(x+w, V)]^2 \\
&= E\left[E\left[Y^2(x+w, V)|V\right]\right] - E[E[Y(x+w, V)|V]]^2
\end{aligned} \tag{7.52}$$

The first term in the fourth line of Eq. (7.58) can be further expanded using the law of total expectation as follows:

$$\begin{aligned}
&E\left[E\left[Y^2(x+w, V)|V\right]\right] \\
&= E[Var[Y(x+w, V)|V]] + E\left[Y[(x+w, V)|V]^2\right] \\
&= E[Var[Y(x+w, V)|V]] + E\left[E[Y(x+w, V)|V]^2\right]
\end{aligned} \tag{7.53}$$

By substituting Eq. (7.53) into Eq. (7.52), one can rewrite Eq. (7.52) as follows: Because

$$\begin{aligned}
\sigma_3^2(f(X)) &= Var[Y(X, V)] \\
&= E[Var[Y(x+w, V)|V]] + E\left[E[Y(x+w, V)|V]^2\right] - E[E[Y(x+w, V)|V]]^2 \\
&= \int_w Var[Y(x+w, V)|V]p(w)dw + \int_w \left(y_{mf}^2(x+w)\right)p(w)dw - \left[\int_w (y_{mf}(x+w))p(w)dw\right]^2
\end{aligned} \tag{7.54}$$

where $Var[Y(x+w,V)|V] = s^2(y_{mf}(x+w))$ is the MSE of the HK prediction, recalling Eq. (7.47), one can obtain

$$Var[y_{mf}(X)] = \int_w y_{mf}^2(x+w)p(w)dw - \left[\int_w (y_{mf}(x+w))p(w)dw\right]^2 \quad (7.55)$$

The variance of the objective function can then be expressed as

$$\sigma_3^2(f(X)) = Var[Y(X,V)] = \int_w s^2(y_{mf}(x+w))p(w)dw + Var[y_{mf}(x+w)] \quad (7.56)$$

The combined effect of the uncertainties of the design variables and the MF surrogate model can be incorporated into the evaluation of the mean and variance of the objective function by adopting Eqs. (7.48) and (7.49) in place of Eqs. (7.46) and (7.47). By substituting Eqs. (7.48) and (7.49) into Eq. (7.42), a robust optimum that considers the uncertainties associated with the design variables and the MF surrogate model can be obtained.

(2) Demonstration: Design of a long cylindrical pressure vessel for compressed natural gas

Let us consider the engineering example of a design optimization problem for a cylindrical pressure vessel. The objective is to minimize the total material consumed for manufacturing. The five continuous design variables and their ranges are listed in Table 7.2. The geometry, model parameters and loading force of the long cylindrical pressure vessel are illustrated in Fig. 7.22.

Other geometric parameters are predefined and remain fixed during the optimization process. The optimization is subject to two design constraints, the maximum allowable stress and the minimum volume. The cylindrical pressure vessel is subjected to a uniformly distributed load of $P = 23$ MPa. The Young's modulus and Poisson's ratio are $E = 207$ GPa and $u = 0.3$, respectively. The maximum allowable stress and the minimum volume are $\sigma_{al} = 250$ MPa and $V_{low} = 0.6\ \text{m}^2$, respectively.

Here, the original deterministic problem (Zhou et al. 2016a) is modified to an RO problem by assuming that two of the design variables, i.e. the inside diameter of

Table 7.2 Ranges of the design variables

Design variables	Range (mm)
The height of the end part h_1	280–320
The inside diameter of the end part r_1	40–50
The thickness of the end part t_1	19–27
The inside diameter of the body part r_2	165–205
The thickness of the body part t_2	13–23

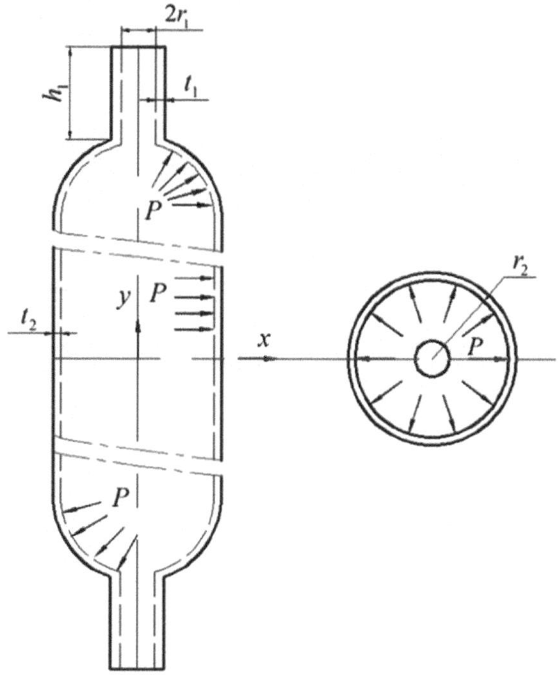

Fig. 7.22 Schematic diagram of the cylindrical pressure vessel

the body and the thickness of the body, both follow normal distributions of $R_2 \sim (r_2, \sigma_1^2), \sigma_1 = 2$, and $T_2 \sim (t_2, \sigma_2^2), \sigma_2 = 0.5$, respectively. These two uncertain variables are assumed to be independent. Then, the modified RO problem can be specified as follows:

$$\begin{aligned} \min: & \ F_{TC}^r = \mu(TC) + 3\sigma(TC) \\ s.t. & \ g_1^r = \mu(\sigma_s) + 3\sigma(\sigma_s) \le \sigma_{al} \\ & \ g_2^r = \mu(V) - 3\sigma(V) \ge V_{low} \end{aligned} \tag{7.57}$$

where V is the volume of the cylindrical pressure vessel and TC is the total amount of material consumed for manufacturing. The quantities V and TC are calculated as follows:

$$V = \pi r_2^2\left(6000 - h_1 - \sqrt{r_2^2 - t_1^2}\right) + 2\pi h_1 r_1^2 + \pi(r_1^2 + r_2^2)\sqrt{r_1^2 - r_2^2} + \frac{1}{3}\pi\left(\sqrt{r_1^2 - r_2^2}\right)^3 \tag{7.58}$$

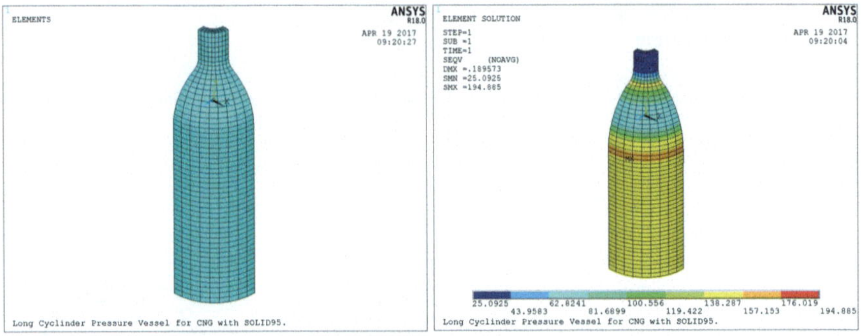

Fig. 7.23 Three-dimensional (3D) HF model of the cylindrical pressure vessel

$$
\begin{aligned}
TC = {} & \pi\left[(t_2+r_2)^2-r_2^2\right]\left(6000-h_1-\sqrt{r_2^2-r_1^2}\right)+2\pi h_1\left[(t_1+r_1)^2-r_1^2\right] \\
& +\pi\left[(t_1+r_1)^2+(t_2+r_2)^2\right]\sqrt{(t_2+r_2)^2-(t_1+r_1)^2}+\frac{1}{3}\pi\left[\sqrt{(t_2+r_2)^2-(t_1+r)^2}\right]^3 \\
& -\pi\left(r_2^2+r_1^2\right)\sqrt{r_2^2-r_1^2}-\left(\sqrt{r_2^2-r_1^2}\right)^3
\end{aligned}
\tag{7.59}
$$

The maximum von Mises stress of the pressure vessel cannot be obtained analytically. Therefore, HK is used to fit the relationship between the stress response and the design variables. In this analysis, ANSYS 18.0 was used as the simulation tool for the stress response. An axially symmetrical three-dimensional (3D) finite element model with hexahedral meshes was selected as the HF model, which is depicted in Fig. 7.23.

Correspondingly, Fig. 7.24 illustrates the one-dimensional (1D) finite element model used as the low-fidelity (LF) model.

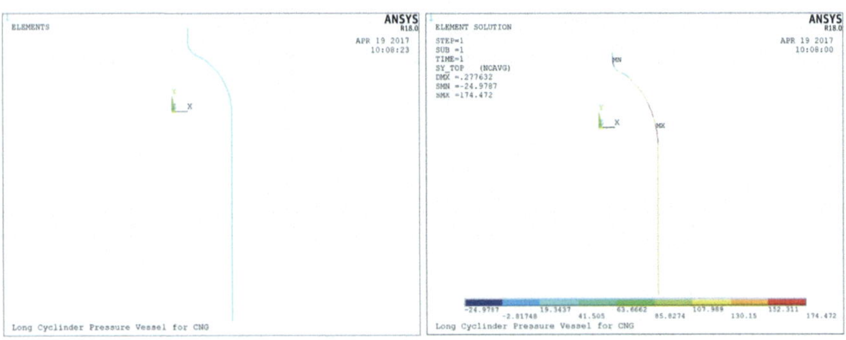

Fig. 7.24 One-dimensional (1D) LF model of the cylindrical pressure vessel

Table 7.3 Deterministic optimization results for the engineering case

Design variables and responses	h_1	r_1	t_1	r_2	t_2	F_{TC}^d	g_1^d	g_2^d
Values	308.95	40.08	19.01	181.74	13.00	0.0949	249.18	0.600

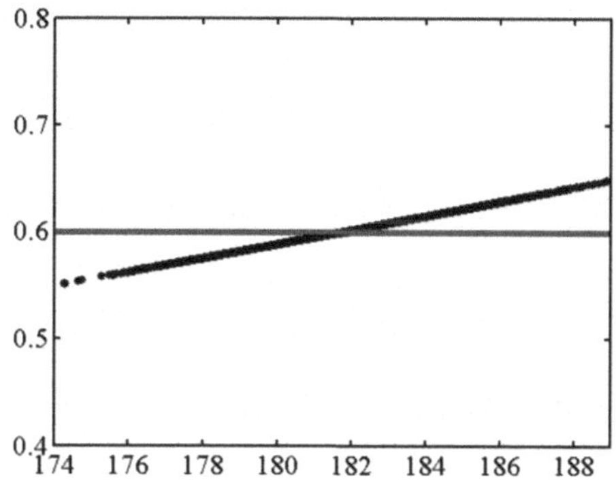

Fig. 7.25 Values of g_2^d with variations in r_2

In this engineering example, 60 sample points were simulated for the LF surrogate model, and the total number of sample points used to develop the HF surrogate model was limited to 15. Deterministic optimization was performed first to illustrate that RO is, in fact, necessary. The deterministic optimization results are summarized in Table 7.3. Table 7.3 shows that at the deterministic optimum, the constraint on g_2^d is active.

Since the constraint on g_2^d is active, when the uncertain variable r_2 varies, the g_2^d constraint can be violated, as shown in Fig. 7.24. Figure 7.24 presents the results of generating 10,000 Monte Carlo samples and calculating the corresponding values of g_2^d. The arrows indicate the feasible direction of the g_2^d constraint. It can be concluded from Fig. 7.25 that the deterministic optimum is not robust and that RO is therefore necessary.

Robust solutions considering only the design variable uncertainty and considering both the design variable and MF surrogate model uncertainties are summarized in Table 7.4.

To validate the actual robustness and feasibility of the solutions, 50 Monte Carlo samples of the random variables in the vicinity of each robust solution were plugged into the 3D HF finite element model. As observed in Table 7.4, although the solution obtained using the robust design approach considering only the design variable uncertainty has a smaller objective value, the actual constraint value ($g_1^r = \mu(\sigma_s) + 3\sigma(\sigma_s) = 266.99$ MPa) calculated from the confirmed points is greater than 250 MPa, meaning that the robust constraint limit ($g_1^r = \mu(\sigma_s) + 3\sigma(\sigma_s) \leq 250$ MPa) is violated. This finding demonstrates that ignoring the uncertainty of the MF surrogate model can result in an infeasible solution. Although the robust optimum found using the MF-surrogate-assisted RO

Table 7.4 Comparison and verification of optima for the example engineering case

Design variables and responses	Only considering design variable uncertainty	Considering both design variable and multi-fidelity surrogate model uncertainties
h_1	303.18	295.77
r_1	50.00	44.00
t_1	19.00	21.16
r_2	205.00	190.30
t_2	19.14	22.06
F^r_{TC}	0.1722	0.1845
$\mu(V)$	0.770	0.658
$\sigma(V)$	0.019	0.013
$g^r_2 = \mu(V) - 3\sigma(V)$	0.713	0.620
$\mu(\sigma_s)$	248.47	205.77
$\sigma(\sigma_s)$	6.17	4.78
$g^r_1 = \mu(\sigma_s) + 3\sigma(\sigma_s)$	266.99	220.10

approach has a larger objective value, the confirmed constraint value achieved using the MF-surrogate-assisted RO approach safely satisfies the constraint limit. This result demonstrates that the MF-surrogate-assisted RO approach may sacrifice the overall objective value to ensure robust solution feasibility.

7.2.2 Surrogate-Model-Based Deterministic RO Approaches

7.2.2.1 Basic Concept of Deterministic RO

A general formulation for an optimization problem with interval uncertainties is given in Eq. (7.60).

$$\begin{aligned} \min \quad & f(\boldsymbol{x},\boldsymbol{p}) \\ \text{s.t.} \quad & g_j(\boldsymbol{x},\boldsymbol{p}) \leq 0 \quad j = 1,2\ldots J \\ & \boldsymbol{x}_{lb} \leq \boldsymbol{x} \leq \boldsymbol{x}_{ub} \\ & \boldsymbol{p}_0 - \Delta\boldsymbol{p}_l \leq \boldsymbol{p} \leq \boldsymbol{p}_0 + \Delta\boldsymbol{p}_u \end{aligned} \tag{7.60}$$

where f is the objective function; $\boldsymbol{x} = (x_1, x_2, \ldots x_N)^T$ is the design variable vector; $\boldsymbol{x}_{lb}$ and $\boldsymbol{x}_{ub}$ are the lower and upper bounds, respectively, on $\boldsymbol{x}$; $\boldsymbol{g} = (g_1, g_2, \ldots g_J)$ are the constraints; and $\boldsymbol{p} = (p_1, p_2, \ldots p_G)^T$ is the parameter vector, which is fixed for a particular optimization run but can exhibit uncertainty. The uncertainties in the parameters $\boldsymbol{p}$ are modelled as intervals, where $\boldsymbol{p}_0$ is the nominal value of $\boldsymbol{p}$ and the upper and lower bounds of the uncertainty region of $\boldsymbol{p}$ are $\Delta\boldsymbol{p}_l$ and $\Delta\boldsymbol{p}_u$, respectively. Let $\Delta\boldsymbol{p}^- = [\Delta p_1^-, \Delta p_2^-, \ldots, \Delta p_G^-]$ and $\Delta\boldsymbol{p}^+ = [\Delta p_1^+, \Delta p_2^+, \ldots, \Delta p_G^+]$ be the lower and upper bounds, respectively, of the variable variations, and let the corresponding normalization of $\Delta\boldsymbol{p}$ be defined as $\Delta\bar{\boldsymbol{p}}_i = \begin{cases} \Delta\boldsymbol{p}_i/\Delta\boldsymbol{p}^-, \Delta\boldsymbol{p}_i \le 0 \\ \Delta\boldsymbol{p}_i/\Delta\boldsymbol{p}^+, \Delta\boldsymbol{p}_i > 0 \end{cases}$. Note that if a design variable is uncertain, then this uncertainty is also modelled as an interval, as in the case of $\boldsymbol{p}$.

The essential goal of RO for Eq. (7.60) is to find an optimum $\boldsymbol{x}$ with certain properties. Some basic concepts are introduced as follows:

(1) Objective robustness: The objective function always varies within its acceptable objective variation range (AOVR) due to perturbation of the uncertain variables in the variable variation range (VVR). In a real-world engineering problem, this AOVR is pre-specified by the decision-makers in accordance with a certain objective robustness requirement.
(2) Feasibility robustness: The constraints are not violated due to perturbation of the uncertain variables, even in the worst-case situation.
(3) Optimality: Given that the abovementioned robustness targets are achieved, the optimum also yields the best objective value.

Referring to the reverse-model-based robust optimization (RMRO) approach presented in Ref. (Gunawan and Azarm 2004), a nested RO approach corresponding to Eq. (7.60) is formulated as follows:

$$\begin{aligned} \min\ & f(\boldsymbol{x},\boldsymbol{p}_0) \\ \text{s.t.}\ & g_j(\boldsymbol{x},\boldsymbol{p}_0) \le 0 \quad j = 1,2\ldots J \\ & 1-\eta_f \le 0 \\ & 1-\eta_g \le 0 \end{aligned} \tag{7.61}$$

where η_f is the objective robustness index and η_g is the feasibility robustness index.

For a given $\boldsymbol{x}_0$, η_f can be obtained by solving the following optimization problem (Gunawan and Azarm 2004):

$$\begin{aligned} \eta_f = \min\ & R_f(\Delta\bar{\boldsymbol{p}})/R_0 = \left[\sum_{i=1}^{G}(\Delta\boldsymbol{p}/\Delta\boldsymbol{p}_T)^2\right]^{\frac{1}{2}}/(G)^{1/2} \\ \text{s.t.}\ & \frac{[\Delta f_0]^2}{[f(\boldsymbol{x}_0,\boldsymbol{p}_0+\Delta\boldsymbol{p}) - f(\boldsymbol{x}_0,\boldsymbol{p}_0)]^2} - 1 = 0 \end{aligned} \tag{7.62}$$

where $\Delta\boldsymbol{p}$ denotes the design variables; $\Delta\boldsymbol{p}_T$ denotes the known bounds of the uncertainty region of $\boldsymbol{p}$; Δf_0 is the AOVR, which represents the acceptable change in the objective function value; R_f denotes the diameter of the objective sensitivity region; R_0 denotes the diameter of the parameter uncertainty region in the

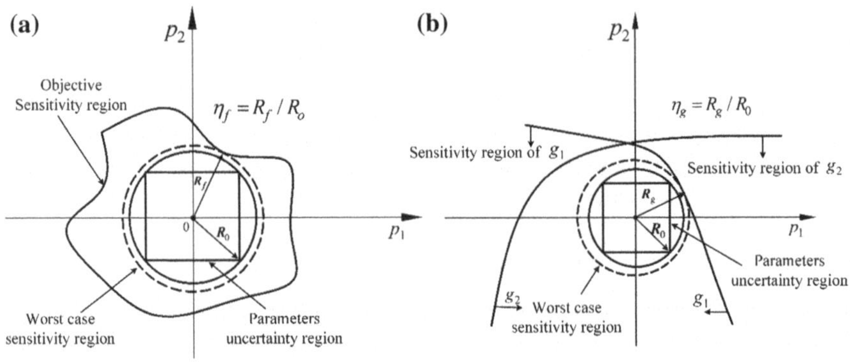

Fig. 7.26 Depiction of the objective robustness index and the feasibility robustness index in two dimensions: **a** objective robustness index η_f, **b** feasibility robustness index η_g

worst-case scenario; and G represents the dimensions of the parameter uncertainty region.

Once η_f is obtained, the relative sizes of the objective sensitivity region and the parameter uncertainty region in the worst-case scenario can be determined, as shown in Fig. 7.26a. The constraint in this problem reflects the fact that a point on the boundary of the sensitivity region must satisfy $[\Delta f_0]^2 = [f(\boldsymbol{x}_0, \boldsymbol{p}_0 + \Delta \boldsymbol{p}) - f(\boldsymbol{x}_0, \boldsymbol{p}_0)]^2$. When $\eta_f \geq 1$, the objective sensitivity region associated with the design alternative covers the parameter uncertainty region, indicating that the design alternative is robust in terms of the objective.

Similarly, for a given $\boldsymbol{x}_0$, η_g can be obtained by solving the following optimization problem (Gunawan 2004):

$$\begin{aligned} &\eta_g = \min R_g(\Delta \bar{\boldsymbol{p}})/R_0 = \left[\sum_{i=1}^{G} (\Delta \boldsymbol{p}/\Delta \boldsymbol{p}_T)^2\right]^{\frac{1}{2}} / (G)^{1/2} \\ &\text{s.t. } \max\left[g_l(\boldsymbol{x}_0, \boldsymbol{p}_0 + \Delta \boldsymbol{p})\right] = 0 \end{aligned} \tag{7.63}$$

where R_g denotes the diameter of the feasibility sensitivity region.

In Eq. (7.63), η_g represents the relative size difference between the feasibility sensitivity region and the parameter uncertainty region in the worst-case scenario, as shown in Fig. 7.26b. When $\eta_g \geq 1$, the feasible sensitivity region associated with the design alternative covers the parameter uncertainty region, indicating that the design alternative is robust in terms of feasibility.

7.2.2.2 A Kriging-Assisted Deterministic RO Approach

(1) The framework and computational effort of the kriging-assisted deterministic RO approach

To improve the effectiveness of RMRO, surrogate models are built for both the objective robustness index and the feasibility robustness index. As a result, the outer–inner nested optimization framework is transformed into a traditional single-level optimization framework:

$$\begin{aligned} \min\quad & f(\boldsymbol{x},\boldsymbol{p}_0) \\ \text{s.t.}\quad & g_j(\boldsymbol{x},\boldsymbol{p}_0) \le 0 \quad j = 1,2\ldots J \\ & 1 - \hat{\eta}_f \ge 0 \\ & 1 - \hat{\eta}_g \ge 0 \end{aligned} \tag{7.64}$$

where $\hat{\eta}_f$ and $\hat{\eta}_g$ are the robustness index and feasibility robustness index, respectively, predicted by the kriging surrogate models.

Figure 7.27 depicts the data flow of this framework, hereafter called K-RMRO. The contents of the dashed boxes represent the process of constructing the surrogate models, where the data flow is indicated by the dotted lines with arrows. Note that unlike traditional constraints, which divide the design space into two continuous, well-defined feasible and infeasible regions, robustness constraints may generate multiple disjoint feasible regions. Therefore, a GA is selected to solve the outer

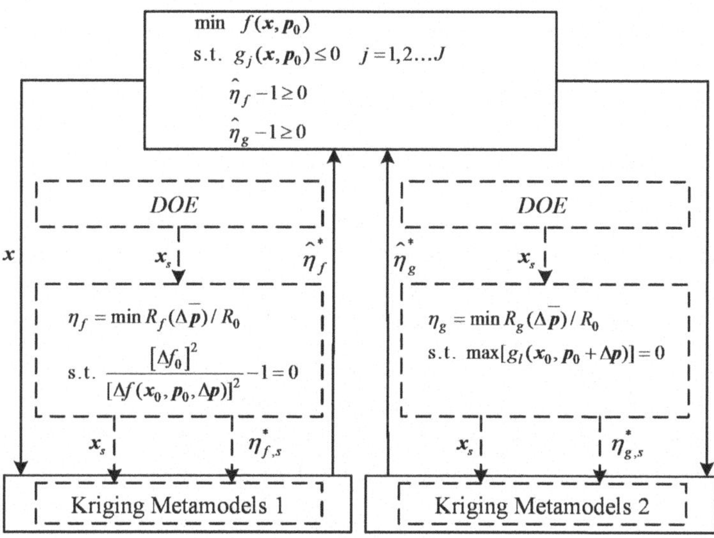

Fig. 7.27 Framework of the K-RMRO approach

problem. The robustness indexes are calculated by employing the subroutine 'fmincon' function in MATLAB. The details of the implementation of K-RMRO are as follows.

Step 1: Generate a set of sample points $\boldsymbol{x}_s$ using Latin hypercube sampling (LHS). $\boldsymbol{x}_s$ will be used as the sample set for generating the kriging surrogate models for the objective robustness index η_f^* and the feasibility robustness index η_g^*.

Step 2: Solve the two inner optimization problems in Eqs. (7.62) and (7.63) for each sample point, thus obtaining the objective robustness index η_f^* and the feasibility robustness index η_g^* corresponding to each sample point.

Step 3: Build kriging surrogate models for the objective robustness index η_f^* and the feasibility robustness index η_g^*.

Step 4: Initialize the outer optimization problem and solve Eq. (7.63) using the GA. During the optimization process, the values of η_f^* and η_g^* for each individual are predicted using the constructed kriging surrogate models. The procedure stops when a prescribed maximum number of iterations is reached.

As mentioned before, the response values obtained from the kriging surrogate models are subject to prediction uncertainties, which may cause K-RMRO to yield false optima. Hence, an improved kriging-assisted RMRO (IK-RMRO) method is presented to enable the consideration of the interpolation uncertainties of the kriging surrogate models. Note that as long as the robustness status of an individual does not change because of the adoption of the kriging surrogate model, the robustness of that individual can be predicted using the kriging surrogate model instead of being judged based on the inner RO problem. However, if the robustness status of the individual does change, then the inner RO problem should be solved. Therefore, an objective switching criterion is introduced in IK-RMRO to determine whether the inner RO problem or the kriging surrogate model replacement should be used to evaluate the robustness of individuals. The switching criterion is presented for the objective robustness; a similar criterion can be used for the feasibility robustness.

In each generation, the objective robustness index of each individual as predicted by the kriging model can be either $\eta_f(x_i) \leq 1$ or $\eta_f(x_i) > 1$. When the effects of the interpolation uncertainties of the kriging surrogate models are considered, there are four possible scenarios for a given individual, as depicted in Fig. 7.28.

- Scenario 1: $\eta_f(x_i) \leq 1$ & $\eta_f(x_i) + 2s(x_i) \leq 1$ (point A)
- Scenario 2: $\eta_f(x_i) \leq 1$ & $\eta_f(x_i) + 2s(x_i) \geq 1$ (point B)
- Scenario 3: $\eta_f(x_i) > 1$ & $\eta_f(x_i) - 2s(x_i) < 1$ (point C)
- Scenario 4: $\eta_f(x_i) > 1$ & $\eta_f(x_i) - 2s(x_i) > 1$ (point D)

On the basis of these scenarios, the individuals in the current generation can be divided into two types:

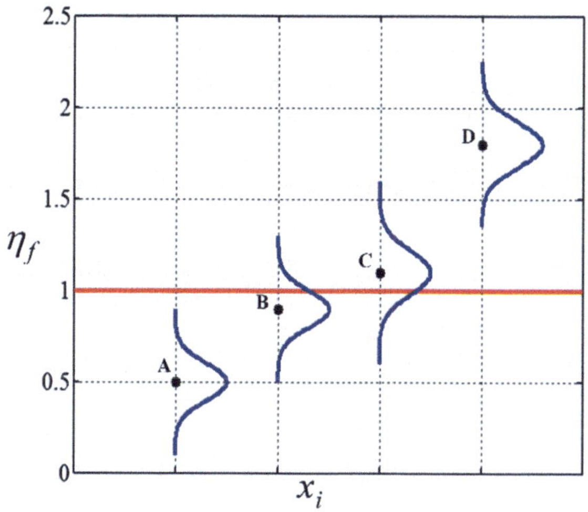

Fig. 7.28 Four different scenarios for candidate designs

Type 1: The objective robustness of these individuals can be predicted using the kriging surrogate model.

Clearly, the prediction error of the kriging surrogate model does not change the robustness status of individuals in Scenarios 1 and 4.

Type 2: The objective robustness of these individuals should be judged based on the inner RO problem.

Clearly, the prediction error of the kriging surrogate model may change the robustness status of individuals in Scenarios 2 and 3.

Figure 7.29 presents the flowchart of the proposed IK-RMRO method. The detailed steps are as follows.

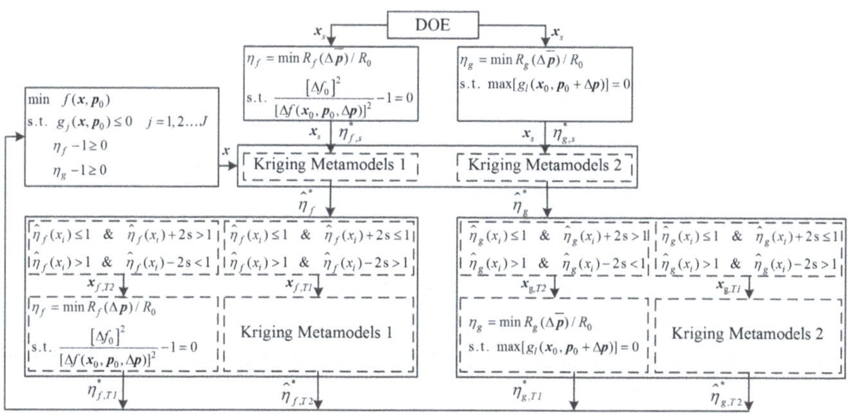

Fig. 7.29 Framework of the IK-RMRO approach

Step 1: Generate a set of sample points $\boldsymbol{x}_s$ using LHS.

Step 2: Solve the two inner optimization problems given in Eq. (7.62) and Eq. (7.63) for each sample point, thus obtaining the objective robustness index η_f^* and the feasibility robustness index η_g^* corresponding to each sample point.

Step 3: Build kriging surrogate models for the objective robustness index η_f^* and the feasibility robustness index η_g^*.

Step 4: Initialize the outer optimization problem. Set the iteration counter to $N = 1$ and generate an initial population $\boldsymbol{x}$ using the GA.

Step 5: Apply the kriging surrogate models to predict the robustness indexes and the corresponding interpolation uncertainty intervals for the current population.

Step 6: Classify the individuals in the current population. This step can be divided into two parts: (1) On the basis of the proposed quantitative criterion for the objective robustness index, divide the current population into two sets $\boldsymbol{x}_{f,T1}$ and $\boldsymbol{x}_{f,T2}$. The objective robustness indexes of individuals in $\boldsymbol{x}_{f,T1}$ will be predicted using the kriging surrogate model, while for the individuals in $\boldsymbol{x}_{f,T2}$, the objective robustness indexes will be obtained by solving the inner RO problem. (2) On the basis of the proposed quantitative criterion for the feasibility robustness index, divide the current population into two sets $\boldsymbol{x}_{g,T1}$ and $\boldsymbol{x}_{g,T2}$. The feasibility robustness indexes of individuals in $\boldsymbol{x}_{g,T1}$ will be predicted using the kriging surrogate model, while for the individuals in $\boldsymbol{x}_{g,T2}$, the feasibility robustness indexes will be obtained by solving the inner RO problem.

Step 7: Calculate the fitness value of each individual in the current population.

Step 8: Generate the next generation based on the genetic operators of selection, crossover and mutation. Set $N = N + 1$.

Step 9: Repeat Steps 5–8. The procedure stops when the prescribed maximum number of iterations $N_{\max}$ is reached.

RMRO can become computationally intractable because the inner problem must be solved for each $\boldsymbol{x}_i$ passed from the outer problem. The superiority of the proposed surrogate-model-assisted RMRO approach relative to the original RMRO approach lies in its computational effort. Below, the required numbers of function calls for RMRO, K-RMRO and IK-RMRO are analysed. Note that a function call refers to the calculation of both the objective and constraint values for a single design point.

Suppose that a GA with a population size P and a maximum number of iterations G is applied to solve the outer problem and that MATLAB's fmincon function is used to solve the inner optimization problem. Then, the number of function calls for the original RMRO approach is

$$FC_1 = P \times G + P \times G \times (Q_1 + Q_2) \tag{7.65}$$

where Q_1 denotes the average number of function calls for the inner problem to obtain the objective robustness index for each candidate and Q_2 is the average number of function calls for the inner problem to obtain the feasibility robustness index for each candidate.

Let the number of sample points used to construct the kriging surrogate models be S; then, the required number of function calls for the K-RMRO approach is

$$FC_2 = P \times G + S \times (Q_1 + Q_2) \tag{7.66}$$

Let P_o and P_g be the numbers of individuals in each generation whose objective robustness and feasibility robustness, respectively, are to be judged based on the inner RO problem. According to the analysis in Sect. 7.3.1, both P_o and P_g must be smaller than the total population size. The required number of function calls for the IK-RMRO approach is

$$FC_3 = P \times G + S \times (Q_1 + Q_2) + P_o \times G \times Q_1 + P_g \times G \times Q_2 \tag{7.67}$$

By comparing the numbers of function evaluations in Eqs. (7.65)–(7.67), it can be found that the RMRO, K-RMRO and IK-RMRO methods are ranked as follows in terms of computational burden: RMRO > IK-RMRO > K-RMRO. A nonlinear numerical example and an engineering case are presented to demonstrate the applicability and efficiency of the IK-RMRO approach. The settings for the GA in the optimization problems are given in Table 7.5. LHS is adopted for sampling points from the design space. The number of sample points S is set to $50T$ in all examples to ensure that the sample points can well reflect the spatial characteristics of the problem, where T is the total number of design variables. To assess the prediction abilities of the constructed kriging surrogate models, two error metrics, the root mean square error (RMSE) and maximum absolute error (MaxAE), are used to measure their global and local accuracies, respectively.

The first example is a two-dimensional nonlinear numerical example adapted from Zhou et al. (2012), and it is used to present a detailed comparison of the RMRO, K-RMRO and IK-RMRO methods. The problem formulation is given as follows:

Table 7.5 GA settings used in the examples

GA settings	Nonlinear numerical example	Pressure vessel design
Population size	40	40
Max. iterations	50	100
Crossover probability	0.95	0.95
Mutation probability	0.05	0.05

Table 7.6 Results comparison for the nonlinear numerical example

Results	Deterministic	RMRO	K-RMRO	IK-RMRO
x_1	−1.826	−1.411	−1.399	−1.447
x_2	0.741	0.268	0.233	0.267
f	−3.287	−1.551	−1.437	−1.567
g_1	−5.919	−6.117	−6.136	−6.166
g_2	0	−1.374	−1.423	−1.360
Function calls	2,000	55,598	4,486	18,558

$$\begin{aligned} &\min f = x_1^3 \sin(x_1+4) + 10x_1^2 + 22x_1 + 5x_1x_2 + 2x_2^2 + 3x_2 + 12 \\ &s.t. \quad g_1 = x_1^2 + 3x_1 - x_1 \sin(x_1) + x_2 - 2.75 \le 0 \\ &\qquad g_2 = -\log_2(0.1x_1 + 0.41) + x_2 e^{-x_1+3x_2-4} + x_2 - 3 \le 0 \\ &\qquad -4 \le x_1 \le 1, \quad -1 \le x_2 \le 1.5 \\ &\qquad \Delta x_1 = \Delta x_2 = 0.4, \Delta f_0 = 2.5 \end{aligned} \tag{7.68}$$

In this example, the design variables x_1 and x_2 have uncertainties of ± 0.4 around their nominal values. The AOVR is $\Delta f = 2.5$. The robust optima obtained via RMRO, K-RMRO and IK-RMRO are listed in Table 7.6. For comparison, the deterministic optimum is also presented. Then, the kriging surrogate models for the objective and feasibility robustness indexes, respectively, have MaxAEs of 0.37 and 0.32 and RMSEs of 0.09 and 0.13. These metrics indicate that the prediction performance of the kriging surrogate models is desirable.

From Table 7.6, it can be observed that the deterministic optimum has the smallest objective value; however, the value of g_2 is equal to 0, which means that the deterministic optimum provides no room for variation relative to the corresponding constraint. Note that although the robust optima obtained via RMRO, K-RMRO and IK-RMRO have objective values larger than that of the deterministic optimum, they provide some amount of 'cushion' for variation relative to the constraint. Thus, these RO approaches may sacrifice the objective value to ensure robustness in terms of the objective and feasibility.

The results listed in Table 7.6 were verified by using LHS to obtain 500 perturbations of x_1 and x_2 around their nominal values within the corresponding ranges Δx_1 and Δx_2 and calculating the $\Delta g_1, \Delta g_2$ and Δf values of the designs corresponding to these perturbed values. To offer improved clarity without losing any important information, Fig. 7.30b, d, f and h shows only the $\max[g_1, g_2]$ value for each perturbation. In Fig. 7.30a, c, e, g, the solid lines represent the acceptable Δf_0 ranges ($\Delta f_0 = \pm 2.5$). In Fig. 7.30b, d, f, h, the solid lines represent the design constraints ($\max[g_1, g_2] = 0$). In these graphs, an optimum is considered robust in terms of the objective if all sample points lie within the solid lines, that is, $\Delta f \le \Delta f_0$, and an optimum is considered robust in terms of feasibility if all sample points lie below the red line, i.e. $\max[g_1, g_2] \le 0$.

It can be seen from Fig. 7.30 that the deterministic optimum becomes infeasible in some cases, while the robust optima of RMRO, K-RMRO and IK-RMRO are always feasible. Considering the objective variations, the objective variations of the deterministic optimum and the robust optima of RMRO and IK-RMRO always remain within the acceptable bounds, while the K-RMRO optimum violates the bounds at some points. In other words, the deterministic optimum does not meet the

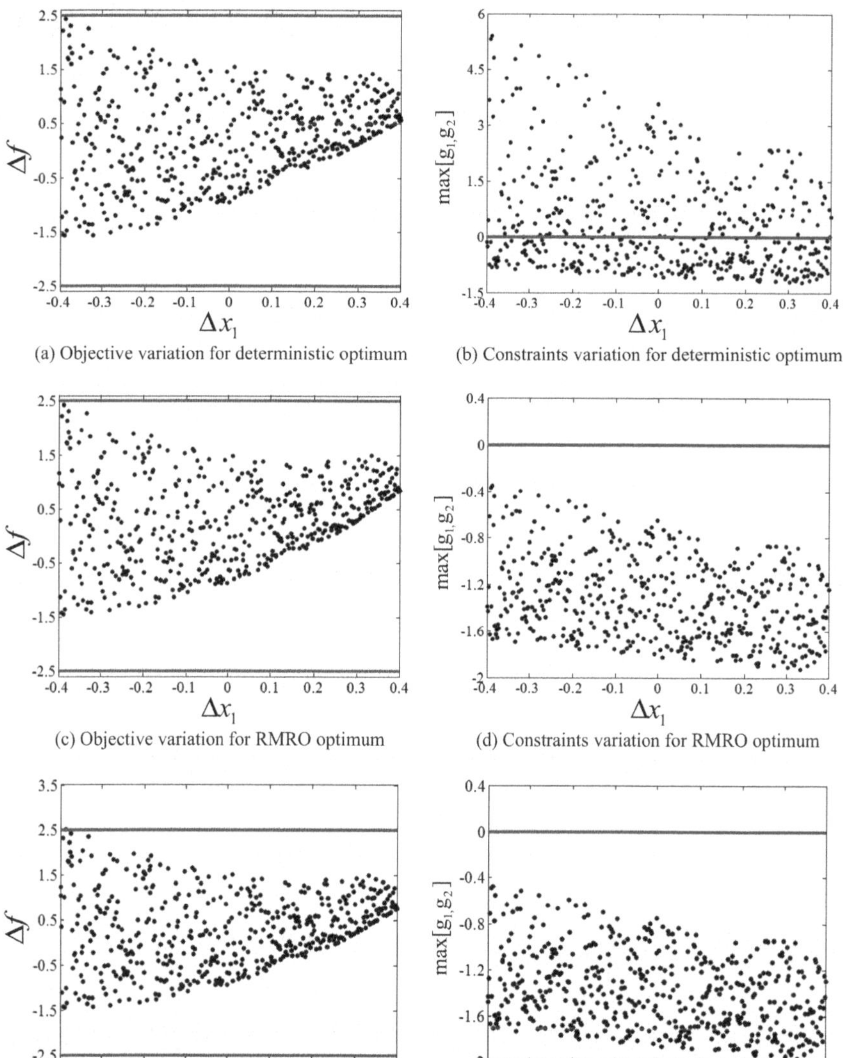

(a) Objective variation for deterministic optimum (b) Constraints variation for deterministic optimum

(c) Objective variation for RMRO optimum (d) Constraints variation for RMRO optimum

(e) Objective variation for K-RMRO optimum (f) Constraints variation for K-RMRO optimum

Fig. 7.30 Robustness verification for the nonlinear numerical example

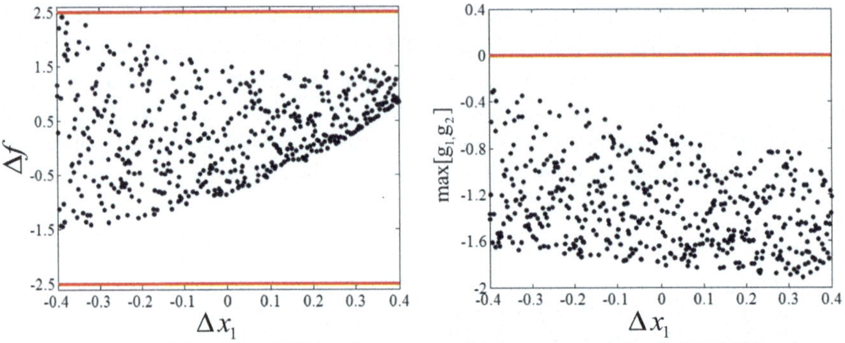

(g) Objective variation for IK-RMRO optimum

(h) Constraints variation for IK-RMRO optimum

Fig. 7.30 (continued)

requirements in terms of the feasibility robustness, and the K-RMRO optimum does not meet the requirements in terms of the objective robustness.

A graphical explanation of these results is presented in Fig. 7.31. In Fig. 7.31, the lines marked with arrows are the constraint boundaries, with the arrows indicating the feasible directions. The ellipses are the contours of the objective function for different optima. The rectangles with dotted lines represent the VVRs. The optima obtained with the different approaches are represented by solid points in Fig. 7.31a–d. As illustrated in Fig. 7.31a, some variable variations cause the deterministic optimum to fall into the infeasible region. In contrast, the VVRs of the RMRO, K-RMRO and IK-RMRO optima lie fully within the feasible region, indicating that these optima are completely robust to variable variations. From Fig. 7.31c, it can be observed that a small part of the VVR of the K-RMRO

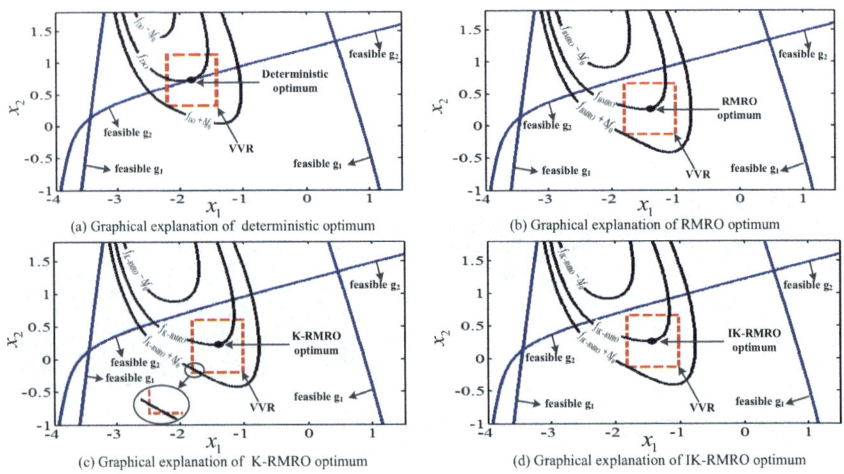

(a) Graphical explanation of deterministic optimum

(b) Graphical explanation of RMRO optimum

(c) Graphical explanation of K-RMRO optimum

(d) Graphical explanation of IK-RMRO optimum

Fig. 7.31 Graphical explanation of the results for the nonlinear numerical example

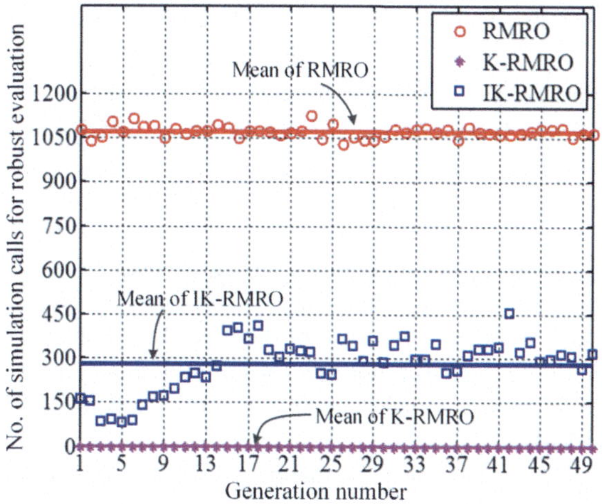

Fig. 7.32 Numbers of simulation calls for robustness evaluations in each generation

optimum lies outside the contours of $f_{K-BMRO} - \Delta f_0$ and $f_{K-BMRO} + \Delta f_0$, while for the other three optima, the VVRs fall completely inside the AOVRs.

Regarding the computational efficiency of each robust method, the number of function calls for K-RMRO is approximately 10 times less than that for RMRO, and the number of function calls for IK-RMRO is approximately 3.5 times less than that for RMRO. Figure 7.32 shows the numbers of simulation calls for robustness evaluations in each generation of the three different robust methods. As shown in Fig. 7.32, K-RMRO does not require robustness evaluations during the RO process, while even the RMRO run with the fewest simulation calls (i.e. 1,029 in generation 26) requires more simulation calls than the IK-RMRO run with the greatest number of simulation calls (i.e. 457 in generation 42). The reason for this difference is that all of the intermediate designs must be evaluated for robustness in RMRO, while only a portion of them must be assessed for robustness in IK-RMRO. Figure 7.33

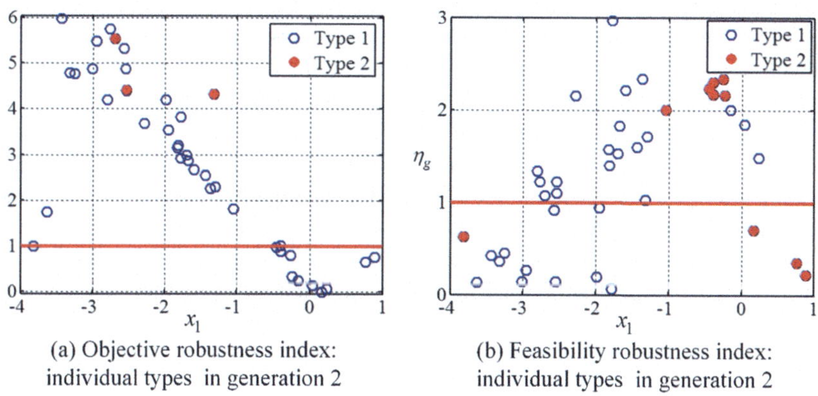

Fig. 7.33 Types of individuals in generation 2 for the IK-RMRO approach

Table 7.7 Results from 30 runs for the nonlinear numerical example

	Deterministic	RMRO	K-RMRO	IK-RMRO
Robustness	0/30	30/30	21/30	30/30
Mean of function calls	2,000	55,765	4,486	19,286

depicts the individual types in generation 2 for IK-RMRO. In Fig. 7.33, the open circles represent individuals whose robustness can be predicted using the kriging surrogate model, while the solid circles represent individuals whose robustness must be judged by means of the inner RO problem. It can be seen from this figure that only a small proportion of the individuals need to be assessed for robustness. As a result, the total number of function calls in IK-RMRO is actually less than that in RMRO.

Due to the stochastic nature of GA, the results obtained might differ from one run to another. To account for this, an additional 29 runs were performed, and the results are summarized in Table 7.7. As shown in Table 7.7, RMRO and IK-RMRO were always able to obtain robust optima, while K-RMRO could guarantee a robust optimum in only 21 of the 30 runs, indicating that it sometimes yields non-robust optima. The deterministic approach did not result in a robust optimum for any of the 30 runs for this example. Regarding the computational cost, the deterministic approach requires the fewest function calls because this method does not require calculating a design's robustness. The levels of computational effort of the kriging surrogate-model-assisted RMRO methods are significantly reduced compared with that of RMRO: K-RMRO requires an average of 4,486 function calls, i.e. 10 times fewer than the 55,765 function calls required by RMRO, and IK-RMRO requires 19,286 function calls, representing a savings of 65.4% function calls compared with RMRO.

The engineering example considered here is a classic pressure vessel design problem, adapted from Kannan and Kremer (1994). Although the problem size is not large for this engineering case, this is a typical and appropriate case for illustrating the applicability and superiority of the IK-RMRO method. The cylindrical vessel is capped at both ends by hemispherical heads, as shown in Fig. 7.34. The objective is to minimize the total cost, including the costs of the material, forming and welding (Coello 2000). Four design variables are considered: x_1 (T_s, the shell thickness), x_2 (T_h, the head thickness), x_3 (R, the inner radius) and x_4 (L, the length of the cylindrical section of the vessel, not including the heads). T_s and T_h are integer multiples of 0.0625 inch (0.15875 cm), which are the available thicknesses of rolled steel plates; R and L are continuous.

Here, the problem is slightly modified by considering the possibility of variations in two of the design variables. The modified RO problem is specified as follows:

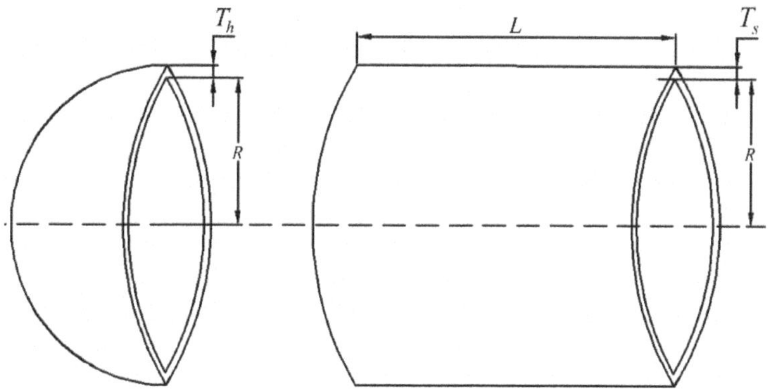

Fig. 7.34 Centre and end sections of the pressure vessel

$$
\begin{aligned}
&\min_{x} f(x) = 0.6224x_1x_3x_4 + 1.7781x_2x_3^2 + 3.1661x_1^2x_4 + 19.84x_1^2x_3 \\
&s.t. \quad g_1(x) = -x_1 + 0.0193x_3 \le 0 \\
&\qquad g_2(x) = -x_2 + 0.0954x_3 \le 0 \\
&\qquad g_3(x) = -\pi x_3^2 x_4 - \frac{4}{3}\pi x_3^3 + 1{,}296{,}000 \le 0 \\
&\qquad g_4(x) = x_4 - 240 \le 0 \\
&where \quad 0 < x_1 < 1.5, 0 < x_2 < 1.5, \quad 30 < x_3 < 50, \ 160 < x_4 < 200 \\
&\qquad \Delta x_1 = 0.01, \Delta x_4 = 0.05, \Delta f_0 = 100
\end{aligned} \tag{7.69}
$$

The detailed results for this RO problem are listed in Table 7.8.

From Table 7.8, it can be observed that the deterministic optimum has the lowest objective value, but the constraints on g_1 and g_2 are active, which indicate that the pre-existing design provides no room for the variables to vary. The relative

Table 7.8 Results comparison for the pressure vessel example

Results	Deterministic	RMRO	K-RMRO	IK-RMRO
$x_1(T_s)$	0.838	0.857	0.831	0.845
$x_2(T_h)$	0.414	0.419	0.415	0.412
$x_3(R)$	43.399	43.453	42.722	43.008
$x_4(L)$	161.158	160.654	169.152	165.758
f	5994.895	6138.735	6035.272	6091.096
g_1	0	−0.018	−0.007	−0.015
g_2	0	−0.005	−0.007	−0.002
g_3	−2.103	−648.742	−527.573	−463.936
g_4	−78.842	−79.346	−70.848	−74.242
Function calls	4,000	50,790	7,675	14,139

difference in the robust optimal objective values between the RMRO and IK-RMRO solutions is only 0.8%, which can be regarded as negligible considering the stochasticity of the GA. In this case, 1000 test points were randomly generated to calculate the values of the MaxAE and RMSE for the constructed kriging surrogate models. The kriging surrogate models for the objective and feasibility robustness indexes, respectively, have MaxAEs of 0.39 and 0.36 and RMSEs of 0.12 and 0.14. These findings illustrate that the prediction capabilities of the kriging surrogate models are desirable.

The optimal results in Table 7.8 were verified by using LHS to generate 500 perturbations of x_1 and x_4 around their nominal values within the corresponding ranges of Δx_1 and Δx_4 and calculating the g_i $(i = 1, 3, 4)$ and Δf_0 values of the designs corresponding to these perturbed values (note that g_2 is independent of x_1 and x_4). To offer improved clarity without losing any important information, Fig. 7.35b, d, f, h shows only the $\max[g_1, g_3, g_4]$ value for each perturbation. In Fig. 7.35a, c, e, g, the solid lines represent the AOVR ($\Delta f_0 = \pm 100$). In Fig. 7.35b, d, f, h, the solid lines represent the design constraints ($\max[g_1, g_3, g_4] = 0$).

From Fig. 7.35, it can be concluded that all four solutions meet the requirements in terms of the objective robustness. However, the deterministic optimum and the K-RMRO optimum become infeasible in some cases, while the robust optima of RMRO and IK-RMRO are always feasible. In other words, the deterministic and K-RMRO optima do not meet the requirements in terms of the feasibility robustness. It should be noted that although there is still space to absorb variations in the objective function for the RMRO and IK-RMRO optima, the objective value cannot be improved because the variances in the constraints have reached their limits.

To further illustrate what happens when x_1 and x_4 vary, the feasible sensitivity region and worst-case sensitivity region were calculated for each optimum in Table 7.8. The feasible sensitivity regions were obtained by constructing the inequality functions $g_i(x_1 + \Delta x_1, x_2, x_3, x_4 + \Delta x_4) \leq 0$, $(i = 1, 3, 4)$ from Eq. (7.42) and then substituting the optimum found with each method into these functions. It can be observed from Table 7.8 that only g_1 and g_3 are active or sensitive to the design variables, while g_4 remains almost the same for each optimum. These observations imply that only g_1 and g_3 make critical contributions to the feasible sensitivity region and that g_4 can be safely dropped from the analysis.

Figure 7.36 presents the robustness information for each optimum. As shown in Fig. 7.36a, the worst-case sensitivity region for the deterministic optimum is very small, approaching a radius of zero. This is because g_1 and g_3 are already almost active at the deterministic optimum, and consequently, little ‘safety margin’ exists for x_1 and x_4 variations (in the worst case). As shown in Fig. 7.36c, the feasible sensitivity region for the K-RMRO optimum does not fully cover the VVR; this finding is attributed to the fact that the interpolation uncertainties of the responses obtained from the kriging surrogate models are ignored. In contrast, the feasible sensitivity regions of the RMRO and IK-RMRO optima are sufficiently large to tolerate more extensive variations in x_1 and x_4. It is also observed that their feasible sensitivity regions completely enclose the VVRs.

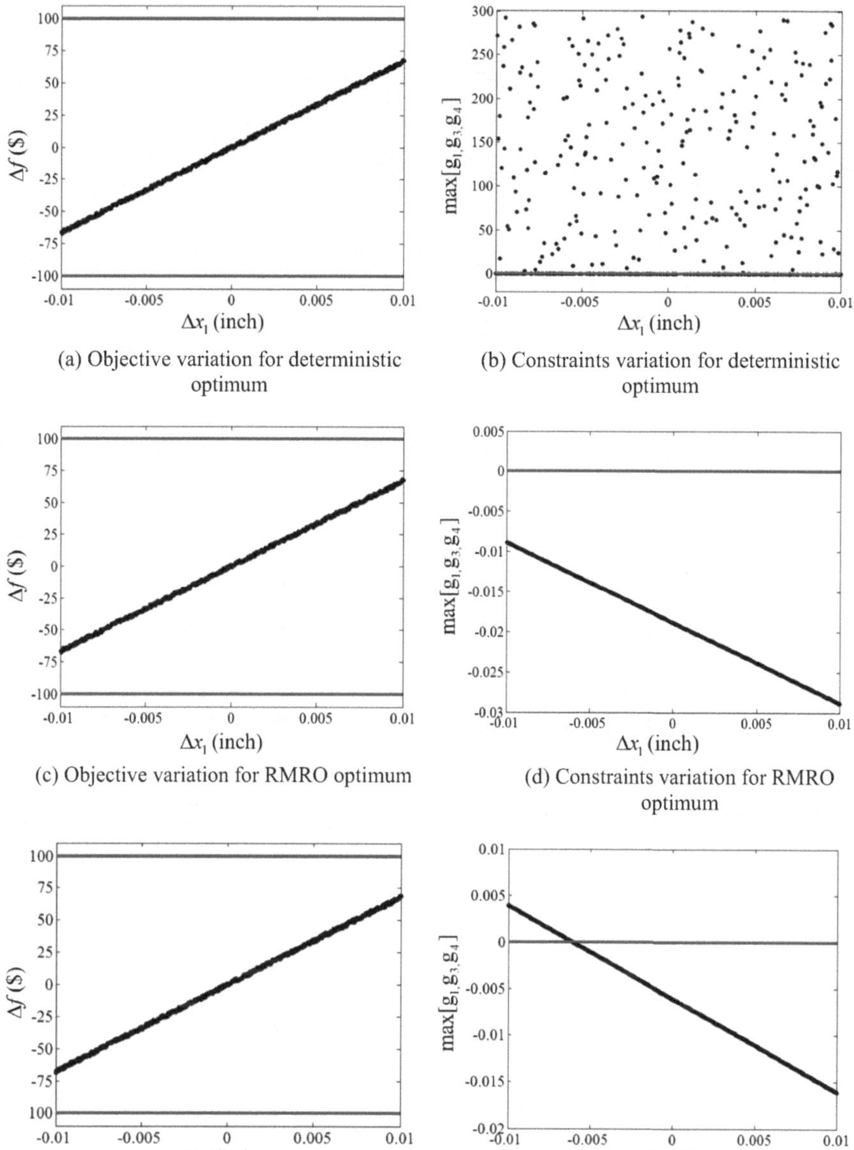

(a) Objective variation for deterministic optimum

(b) Constraints variation for deterministic optimum

(c) Objective variation for RMRO optimum

(d) Constraints variation for RMRO optimum

(e) Objective variation for K-RMRO optimum

(f) Constraints variation for K-RMRO optimum

Fig. 7.35 Robustness verification for the pressure vessel example

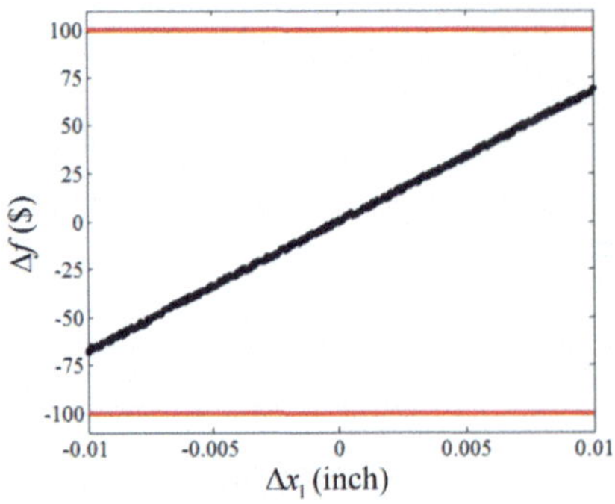

(g) Objective variation for IK-RMRO optimum

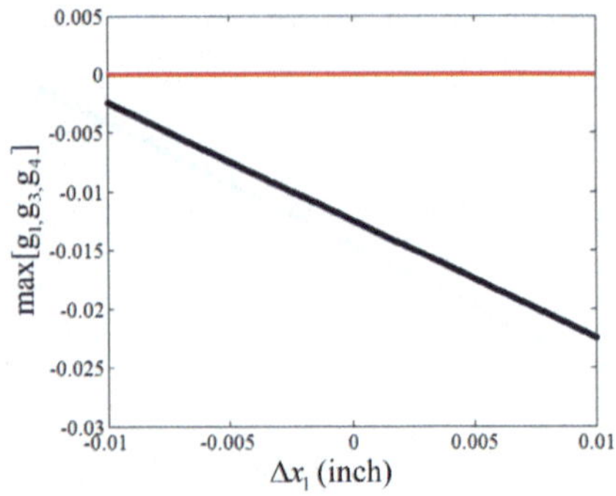

(h) Constraints variation for IK-RMRO optimum

Fig. 7.35 (continued)

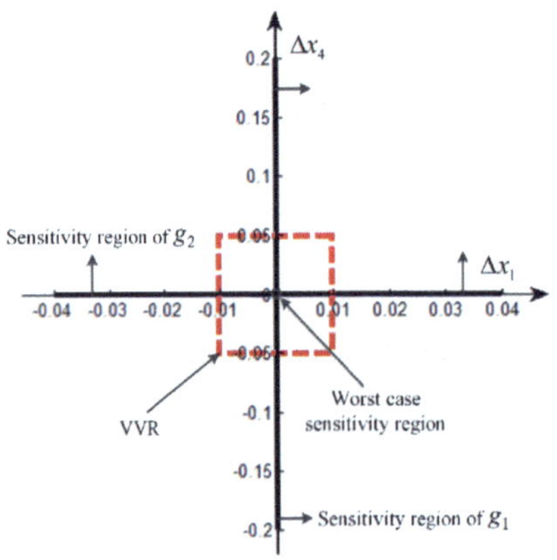

(a) SR and WCSR of deterministic optimum

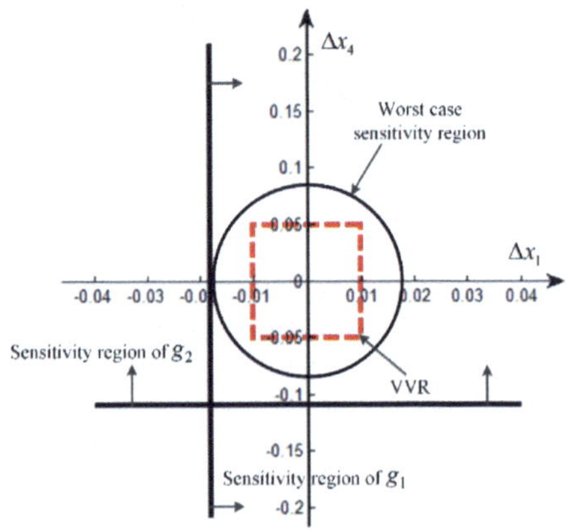

(b) SR and WCSR of RMRO optimum

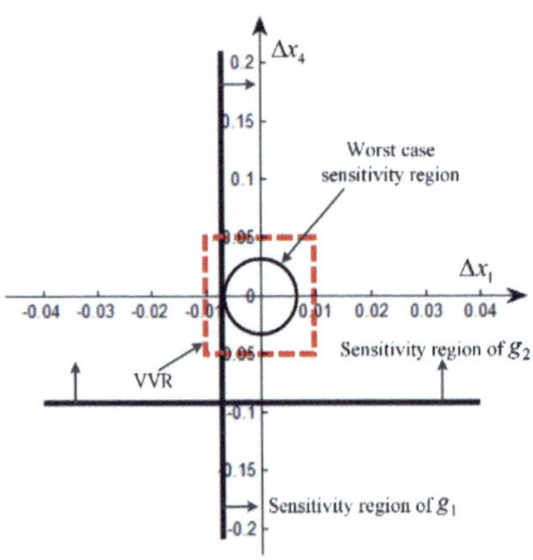

(c) SR and WCSR of K-RMRO optimum

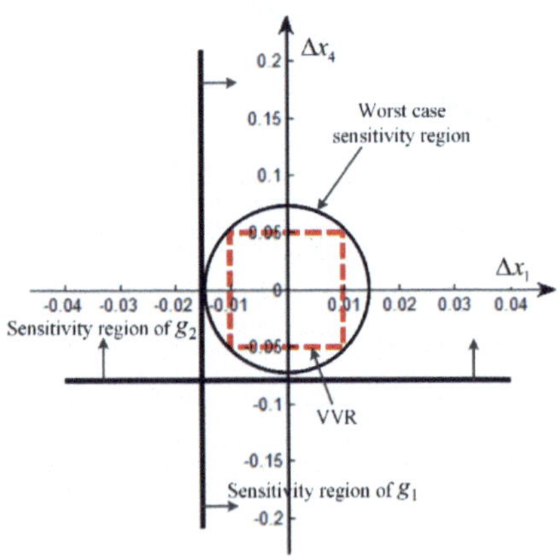

(d) SR and WCSR of IK-RMRO optimum

Fig. 7.36 Robustness information on the results for the pressure vessel design example

Table 7.9 Results from 30 runs for the pressure vessel example

	Deterministic	RMRO	K-RMRO	IK-RMRO
Robustness	0/30	30/30	8/30	30/30
Mean of function calls	4,000	51,146	7,675	13,738

Table 7.9 summarizes the results from a total of 30 runs for this pressure vessel design example. As seen in Table 7.9, both IK-RMRO and RMRO are able to obtain robust optima, and the average number of function calls required by IK-RMRO is 13,738, which is approximately 28% of the function calls required by RMRO. Although K-RMRO can reduce the number of function calls by 44% compared with IK-RMRO, it cannot always guarantee the robustness of the solution. In this analysis, K-RMRO could guarantee a robust optimum in only 8 of the 30 runs.

7.3 Surrogate-Model-Based Evolutionary Optimization

Evolutionary optimization algorithms require a large number of function evaluations to converge to globally optimal or near-optimal solutions (Sun et al. 2013; Cheng et al. 2015b). This requirement has somewhat limited their ability to solve real-world engineering design problems relying on computationally expensive simulation models, which we refer to as HF models. There are three common strategies to improve the efficiency of such algorithms. The first strategy is referred to as fitness inheritance, in which the number of fitness evaluations is reduced by estimating the fitness values of offspring individuals based on their parents (Chen et al. 2002; Bui et al. 2005). The parents are individuals from the previous generation from which the offspring individuals were generated. The second strategy is referred to as fitness imitation. Unlike in fitness inheritance, the fitness values of offspring individuals are estimated based on several selected representative individuals in the current generation (Kodiyalam et al. 1996; Jin 2005). The last strategy is referred to as fitness approximation (using surrogate-model-based methods), in which surrogate models, e.g. kriging (Li et al. 2009a; Li 2011; Liu and Collette 2014), radial basis function (RBF) (Chen et al. 2012; Regis 2013b; Datta and Regis 2016; Sun et al. 2017), neural network (NN) (Song et al. 2012), support vector regression (SVR) (Andrés et al. 2012) or quadratic polynomial fitting (QPF) (Goel et al. 2007) models, are constructed to replace the fitness evaluations to reduce the number of fitness calculations. Since the fitness approximation strategy can lead to the best performance and yields the most efficient methods among the above three strategies, these approaches have attracted widespread interest (Jin 2011; Zhu et al. 2014).

In the broadest sense, surrogate-model-based methods can be divided into two distinct modes: offline mode and online mode (Wang et al. 2016). In the offline mode, a pre-specified number of sample points are used to build a surrogate model, which is subsequently used in place of the simulation model in the evolutionary computations (Chung and Alonso 2004; Lian and Liou 2005; Mogilicharla et al. 2015). The main shortcoming of the offline mode is that it is difficult to reduce the number of fitness evaluations while simultaneously obtaining a surrogate model of the desired accuracy (Li 2011). By contrast, in the online mode, an initial surrogate model is first generated and then adaptively updated following certain model updating strategies throughout the optimization process (Li et al. 2009a; Hamdaoui et al. 2015). Compared with the offline mode, the online mode, which can exploit data from previous iterations, has been reported to be more efficient for evolutionary algorithms (Li 2011; Shimoyama et al. 2013).

In engineering design optimization, there may be multiple conflicting design objectives. Thus, the ability to rapidly understand the trade-offs between multiple conflicting objectives is important (Liu and Collette 2014). A general formulation for a multi-objective optimization problem is given below:

$$\begin{aligned} \min_{x} \ & F(\boldsymbol{x}) = \{f_1(\boldsymbol{x}), f_2(\boldsymbol{x}), \ldots, f_i(\boldsymbol{x}), \ldots, f_M(\boldsymbol{x})\} \\ \text{s.t.} \ & g_j(\boldsymbol{x}) \leq 0, \quad j = 1, 2 \ldots J \\ & \boldsymbol{x}_{lb} \leq \boldsymbol{x} \leq \boldsymbol{x}_{ub} \end{aligned} \tag{7.70}$$

where $F(\boldsymbol{x})$ is the objective function vector, which contains at least two conflicting objective functions; $\boldsymbol{x} = (x_1, x_2, \ldots x_N)^T$ is the design variable vector; $\boldsymbol{x}_{lb}$ and $\boldsymbol{x}_{ub}$ are the lower and upper bounds, respectively, on $\boldsymbol{x}$; and $\boldsymbol{g} = (g_1, g_2, \ldots g_J)$ are the constraints. Since trade-offs exist among the objective functions, the optimization problem expressed in Eq. (7.70) generally has a set of Pareto optimal solutions, that is, there is no optimum that is superior to all other designs in terms of all objectives (Shan and Wang 2005). These solutions are called the Pareto set or Pareto frontier. Since multi-objective evolutionary algorithms (MOEAs) are the main means of solving such problems, considerable attention has been paid to the possibility of adopting the fitness approximation strategy to reduce the cost of these algorithms. In this chapter, several typical online surrogate-model-based multi-objective evolutionary optimization methods are introduced in detail.

7.3.1 A Kriging-Model-Assisted Multi-objective Genetic Algorithm (K-MOGA)

The core factor affecting the success of online surrogate-model-based evolutionary algorithms is the model updating strategy. The most direct model updating strategy is to update the surrogate model by evaluating the individuals with the best fitness values (Jin et al. 2002), the individuals with the largest uncertainty (Branke and Schmidt 2005; Wang et al. 2017) or the individuals that offer a trade-off between improving the surrogate model accuracy and searching for the best fitness values (Jeong et al. 2006; Hu et al. 2008; Hamdaoui et al. 2015). The random selection of individuals to be evaluated using the original fitness function in each generation has also been studied as a strategy for updating the surrogate model (Jin et al. 2000). Preliminary efforts have demonstrated that these updating strategies with a predefined number of updates may cause oscillation because the accuracy of the surrogate model may fluctuate significantly during the optimization process (Ratle 1998). To address this issue, Li (Li et al. 2007) proposed a kriging surrogate-model-assisted multi-objective genetic algorithm (K-MOGA), in which an objective criterion is introduced to select individuals for simulation based on whether the dominance statuses of the individuals (according to the original function) change due to the uncertainty of the surrogate model.

The conventional multi-objective genetic algorithm (MOGA) used here is based on NSGA (Kalyanmoy 2001) combined with an elitism strategy. In each generation of the conventional MOGA, the current population is composed of two parts, namely, non-dominated points and dominated points, and the response values at the points in the initial population are calculated using a simulation model. When a kriging model is used in place of the simulation model, the response values predicted by the kriging model are subject to a prediction error. If the prediction error indicates that the dominance statuses of the design points in the current generation will not change because of the use of the kriging model, it is acceptable to use the kriging model instead of the simulation model. If the dominance status does change, then the design points that are predicted to contribute to this change are observed (i.e. their objective function values are computed using the simulation model); otherwise, the kriging model is used to obtain all response values.

A quantitative measure of dominance, the minimum of minimum distance (MMD), is used as the basis of the criterion for determining whether the values predicted by the kriging model should be accepted. The MMD is defined as the minimum distance between all pairs of non-dominated and dominated points in the objective space, and it is calculated as follows. First, the individuals in the current population are divided into two sets: the non-dominated set and the dominated set. Then, the MMD is projected into each dimension in the objective space to obtain the MMD_{f_m} (m = 1, …, M), as shown in Fig. 7.37.

The prediction error at a design point x is defined as

$$I_m(\boldsymbol{x}) = \sigma_m(\boldsymbol{x}), m = 1, \ldots, M \tag{7.71}$$

where $\sigma_m(\boldsymbol{x})$ is the standard deviation of the predictions of the kriging model.

Figure 7.38 depicts the relation between the prediction interval and the *MMD*. As shown in Fig. 7.38a, the dominance statuses of the individuals will not change

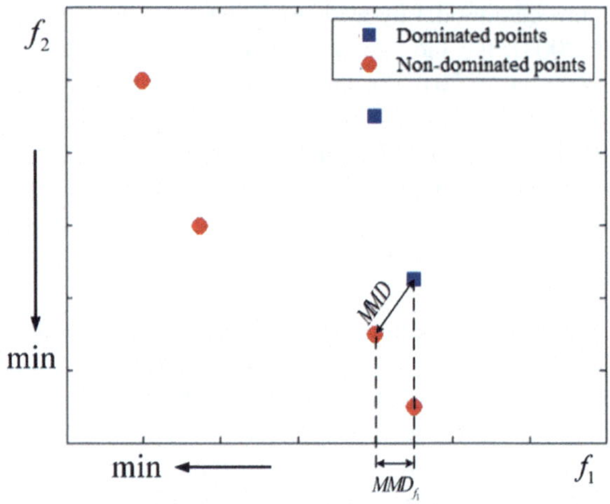

Fig. 7.37 MMD and MMD_{f_m} in the objective space

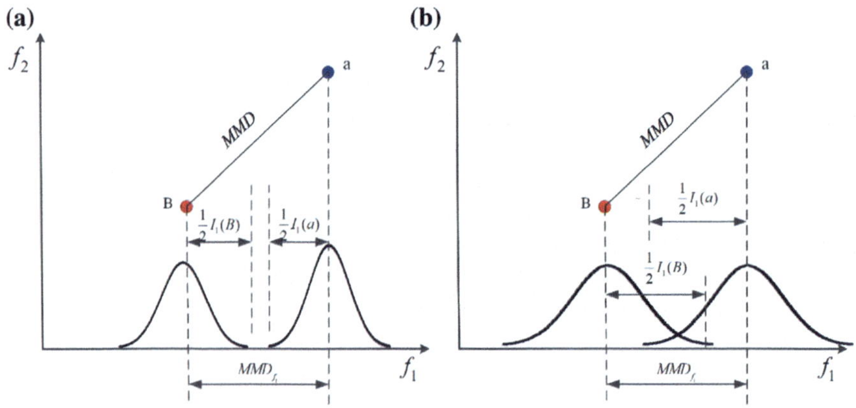

Fig. 7.38 Relation between the prediction interval and the *MMD*

when the relation between the MMD and the prediction interval satisfies the following condition:

$$I_m(B) + I_m(a) \leq 2MMD_{f_m} \tag{7.72}$$

Otherwise, the dominance statuses of individuals may change, as shown in Fig. 7.38b.

The steps of K-MOGA are as follows:

Step 1. Initialization. Start by generating an initial population. Simulation models are called to calculate the responses for the individuals. The initial kriging models are constructed based on the initial samples.
Step 2. Generate a new population by means of GA operations.
Step 3. Apply the current kriging models to predict the response values for the current population. Obtain the non-dominated set and the dominated set. For individuals that do not satisfy Eq. (7.72), the simulation model is used to calculate their responses.
Step 4. Calculate the fitness value of each point.
Step 5. Identify the non-dominated points and update the kriging models. The individuals whose response values have been calculated through simulation are used to update the kriging models.
Step 6. Check the stopping criteria. If the stopping criteria are satisfied, the algorithm terminates; otherwise, it continues.
Step 7. Generate the next population. Return to Step 2.

7.3.2 A Multi-objective Variable-Fidelity Optimization Method for GAs

The intrinsic drawbacks of single-fidelity surrogate modelling, i.e. the time-consuming nature of running HF simulation models and the risk that incorporating inexpensive LF models directly into the MOGA may result in an inaccurate Pareto frontier, have not been solved in the method discussed above. A promising way to achieve a trade-off between high accuracy and high efficiency is to adopt the multi-fidelity modelling (MFM) approach (Zhou et al. 2015a). Although MFM has already been applied in engineering design optimization (Gano et al. 2006b; Huang et al. 2006a; Han et al. 2010), researchers have mainly utilized MFM to model the responses of engineering systems. By contrast, Zhu et al. (2013) have proposed a multi-objective variable-fidelity optimization method for GAs. In Zhu's method, the NSGA-II multi-objective GA optimizer proposed by Deb et al. (Deb et al. 2002) is used to drive the optimization problem. The global–local approximation approach is used in this method, in which the HF function is approximated by a global LF engineering model and a correction factor (Haftka 1991).

In this method, a fixed updating strategy is applied in which the total number and distribution of HF calls are set at the outset of the optimization process. The following procedure is used to distribute these HF calls.

(1) Initially, only the LF model is run for a number of generations to allow a rough Pareto front to evolve from the randomly selected individuals that form the first generation.

(2) After a given number of generations, called the offset, a number of individuals chosen from the current best non-dominated front in the optimizer are sent for HF analysis. These individuals are selected based on a simple inter-individual distance metric in the objective function space. The number of chosen individuals is set equal to an integer multiple of the processing capacity of the computer cluster that is available for solving the problem. This number is referred to as the density of the update. This strategy ensures that 100% of the available processing power is always used when updating the surrogate model. A kriging model is formed based on these data.
(3) Subsequent objective function evaluations use either the LF solution scaled to the kriging model or simply the LF solution, depending on the error of the kriging model at the point at which the objective function is to be evaluated.
(4) After a fixed number of generations have passed (referred to as the spacing), another set of individuals (where the number of individuals is again equal to the density) is selected from the current best non-dominated front and sent for HF analysis. These individuals are selected by computing the prediction errors of the current kriging model at all points in the current best non-dominated front and sending the individuals with the highest errors for HF analysis. A new kriging model is then formed by combining the existing and new HF results.

7.3.3 An Online Multi-fidelity Surrogate-Model-Assisted Multi-objective Genetic Algorithm (OLVFM-MOGA)

In Zhu's method, the cost of the LF model is assumed to be zero. However, this method may lead to a large computational burden when the computational cost of the LF model cannot be ignored. Shu et al. (2018) proposed an online MF surrogate-model-assisted MOGA considering the possible change in status between dominated individuals and non-dominated individuals and the computational costs of the HF and LF models. This method is called OLVFM-MOGA by the authors.

A commonly used MF surrogate model approach based on an additive scaling function is used here, in which the MF surrogate model is obtained by using a scaling function to tune the LF model in accordance with the response values of the HF model. The MF surrogate model can be expressed as

$$\hat{f}_{mf}(\boldsymbol{x}) = \hat{f}_l(\boldsymbol{x}) + \hat{C}(\boldsymbol{x}) \tag{7.73}$$

where $\hat{f}_{mf}(\boldsymbol{x})$ is the MF surrogate model, $\hat{f}_l(\boldsymbol{x})$ represents the LF surrogate model and $\hat{C}(\boldsymbol{x})$ is the scaling function. Kriging models are used to construct the LF surrogate model and the scaling function surrogate model. A kriging model can provide an estimate of the prediction error at an unobserved point. The standard deviation of the predictions of the MF surrogate model can be expressed as

$$\sigma_{mf}(\boldsymbol{x}) = \sqrt{\sigma_l^2(\boldsymbol{x}) + \sigma_C^2(\boldsymbol{x})} \tag{7.74}$$

where $\sigma_l(\boldsymbol{x})$ and $\sigma_C(\boldsymbol{x})$ are the standard deviations of the LF surrogate model and the scaling function surrogate model, respectively, which can be obtained from the corresponding kriging models. The prediction interval at design point $\boldsymbol{x}$ can be defined as

$$\begin{cases} I(\boldsymbol{x}) = c\sigma_{mf}(\boldsymbol{x}) \\ f(\boldsymbol{x}) = \hat{f}_{mf}(\boldsymbol{x}) \pm 0.5I(\boldsymbol{x}) \end{cases} \tag{7.75}$$

where c reflects the confidence level and $I(\boldsymbol{x})$ is the prediction interval. In accordance with the six-sigma criterion used in engineering design (Koch et al. 2004), the value of c is set to 6 in this analysis, representing a confidence level of 99.87% that the true response lies within the prediction interval.

A novel model updating strategy is adopted in this approach to consider the interpolation uncertainty introduced by MFM. In OLVFM-MOGA, the MF model will be updated after each generation by sending individuals for LF/HF analysis in accordance with the model updating strategy. Here, the individual-based updating strategy proposed by Li (2007) is extended to MFM scenarios to consider the computational costs of models of different fidelities.

The MMD, which refers to the minimum distance among all distances between all pairs of non-dominated and dominated points, was introduced in Sect. 7.1 to determine the lower bound on the distance between any two design points such that one of the points is in the non-dominated set while the other is not. The MMD is projected into each dimension in the objective space to obtain the MMD_{f_m} (m = 1, …, M), as shown in Fig. 7.37.

Figure 7.38 depicts the relation between the prediction interval and the *MMD*. As shown in Fig. 7.38a, the dominance statuses of the individuals will not change when the relation between the MMD and the prediction interval satisfies the following condition:

$$I_p(B) + I_p(a) \leq 2MMD_{f_1} \tag{7.76}$$

Otherwise, the dominance statuses of individuals may change, as shown in Fig. 7.38b. To avoid this situation, the prediction intervals for such individuals should be reduced. According to Eqs. (7.74) and (7.75), the prediction intervals can be reduced by sending the individuals for LF or HF evaluations. Considering that the computational burden of the LF model is much less than that of the HF model, the individuals whose dominance statuses may change will first be sent for LF

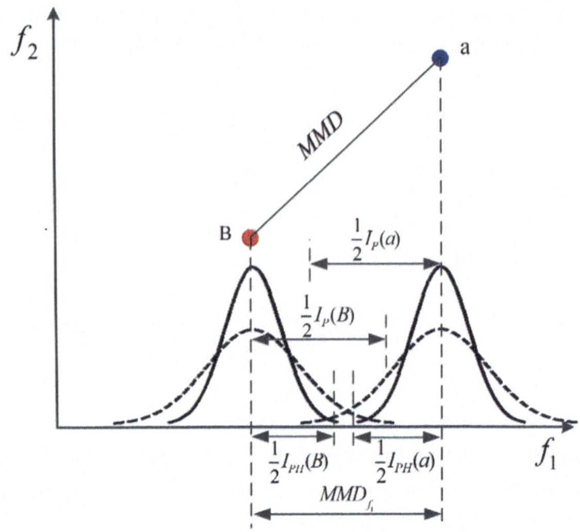

Fig. 7.39 Prediction intervals for individuals after LF evaluations

evaluations. As shown in Fig. 7.39, $I_{PH}(B)$ and $I_{PH}(a)$ denote the prediction intervals after the individuals are sent for LF evaluations. Note that $I_{PH}(B)$ and $I_{PH}(a)$ depend only on the uncertainty of the surrogate model scaling function and that they are smaller than the previous prediction intervals. Then, the relation between the prediction interval and the MMD will be checked based on Eq. (7.76). If it is found that the dominance statuses of the individuals will not change, then the individuals will not be sent for HF evaluations; otherwise, the prediction intervals will be further reduced by sending the individuals for HF evaluations.

The steps of selecting individuals for LF/HF evaluations are listed in Algorithm 7.7. To prevent a very small projection distance MMD_{f_1} from causing an unnecessary computational burden, a very small threshold value ε is set such that when MMD_{f_1} is smaller than ε, the corresponding individuals will not be selected for simulation analysis.

For optimization problems with constraints, the objectives are penalized using a penalty function (Coello 2000) if the individuals are located in an infeasible part of the search space. This will cause such individuals to be dominated and far away from the non-dominated individuals in the output space. Hence, the dominance statuses of these individuals will not change, and they will not be selected for LF or HF simulations based on the proposed updating strategy. Through the iterative process of MOGA, these individuals will be weeded out.

Algorithm 7.7 Algorithm of the updating strategy

Input: the individuals of the current generation

Begin	
$\{x_A, x_B, \ldots\}_{\text{non-dominated}}$, $\{x_a, x_b, \ldots\}_{\text{dominated}}$	←Obtain the non-dominated and dominated points using NSGA-II.
$MMD = \min_{\substack{p = x_A, x_B, \ldots \\ q = x_a, x_b, \ldots}} \{\lVert f(p) - f(q) \rVert_2\}$	←Calculate the MMD between non-dominated and dominated points.
MMD_{f_1}	←Obtain MMD_{f_1} by projecting the MMD along the objective function axis.
if $MMD_{f_1} > \varepsilon$	
$I_P(x_A), I_P(x_B), \ldots, I_P(x_a), I_P(x_b), \ldots$	←Calculate the prediction intervals at the points by means of kriging model prediction.
$I_P(x_1) \geq I_P(x_2) \geq \ldots \geq I_P(x_n) \geq 0$	←Sort the individuals in descending order of their prediction interval sizes.
For $i = 1$ **to** n **do**	
if $I_P(x_i) + I_P(x_{i+1}) > 2MMD_{f_1}$	
$\text{LF}(x_i)$	←Send x_i for LF analysis.
if $I_{PH}(x_i) + I_{PH}(x_{i+1}) > 2MMD_{f_1}$	
$\text{HF}(x_i)$	←Send x_i for HF analysis.
end if	
else if	
Break for	
end if	
end for	

Output: the individuals sent for LF/HF analysis and the corresponding responses

The detailed steps are described as follows.
Begin

Step 1: Select the initial LF and HF sample points and obtain the corresponding responses to construct the initial variable-fidelity surrogate model (VFM) (the responses for each objective will be normalized).
Step 2: Initialize the population of NSGA-II.
Step 3: Initialize the generation counter at $N = 1$ and evaluate the fitness values of the individuals using the constructed VFM.
Step 4: Obtain the non-dominated set and the dominated set.
Step 5: Select new HF and LF sample points to update the VFM using Algorithm 1.
Step 6: Generate the new population and update the generation counter to $N = N + 1$.
Step 7: Evaluate the fitness values of all individuals using the updated VFM.
Step 8: Check whether the stopping criterion is satisfied. If yes, proceed to Step 9; otherwise, return to Step 4.
Step 9: Output the obtained optimum.

End

7.3.4 Examples and Results

In this section, six numerical examples (ZDT1, ZDT2, ZDT3, FON, POL and QV) adapted from Li et al. (Li 2011) and Liu et al.(Liu and Collette 2014) and an engineering case adapted from Park et al. (Park and Dang 2010) are used to demonstrate the applicability and efficiency of the proposed OLVFM-MOGA approach. In these numerical examples, the LF models are modified versions of the original numerical functions, and the computational cost of an HF sample is assumed to be four times that of an LF sample. For comparison, these examples are also solved using four other methods: (1) a MOGA with the LF model, (2) a MOGA with the HF model, (3) a surrogate model-assisted MOGA with the individual-based updating strategy (K-MOGA) and (4) Zhu's method (Zhu et al. 2013).

The maximum number of generations was set to 150 for the QV problem and 100 for other problems. The settings for the other parameters of the various methods are summarized in Table 7.10. Optimal Latin hypercube design (OLHD) (McKay et al. 2000) was adopted for generating the initial points. For the numerical examples, we solved each problem 30 times with each method to account for the influence of randomness. For the engineering case, the optimization problem was

Table 7.10 Algorithm parameter settings for all examples

Parameter	OLVFM-MOGA	IUK-MOGA	MOGA with LF model	MOGA with HF model
Population size	40	40	40	40
Initial LF samples	60	/	/	/
Initial HF samples	20	40	/	/
ε	1e−3	/	/	/

solved 15 times each with K-MOGA, Zhu's method and OLVFM-MOGA but only one time for the MOGA with the HF model because of the large time cost.

Quality metrics for the Pareto optima

Two metrics proposed in the literature (Wu and Azarm 2001; Cheng et al. 2015c), i.e. the relative hyperarea difference (RHD) and the overall spread (OS), were calculated to measure the convergence quality and diversity of the Pareto optima. Figure 7.40 depicts the geometrical interpretations of these two metrics for a two-dimensional case. Let the current robust Pareto set be $P = \{a, b, c, d\}$, and let p_{good} and p_{bad} denote the extreme good and bad points, respectively. The RHD, as shown in Fig. 7.40a, is defined as the relative difference between the area bounded by p_{good} and p_{bad} and the area between p_{bad} and the current Pareto set P.

$$RHD = \frac{HA(p_{bad}, p_{good}) - HA(p_{bad}, a, b, c, d)}{HA(p_{bad}, p_{good})} \tag{7.77}$$

The OS, as shown in Fig. 7.40b, is defined as the ratio between the area bounded by the two extreme points of the current Pareto set P and the area bounded by p_{good} and p_{bad}.

$$OS = \frac{HA[extremes(P)]}{HA(p_{bad}, p_{good})} \tag{7.78}$$

The RHD and OS represent the convergence quality and diversity, respectively, of the obtained Pareto frontier. The smaller the value of the RHD is, the higher the convergence of the Pareto frontier, while a larger value of the OS indicates greater diversity of the Pareto frontier. The settings of P_{good} and P_{bad} for the different examples are listed in Table 7.11.

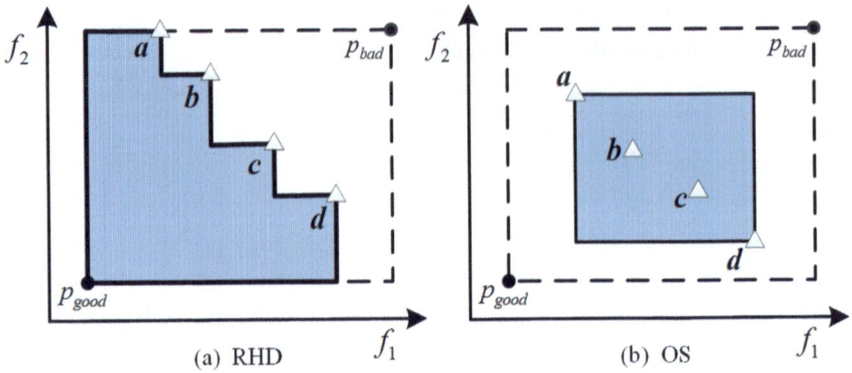

Fig. 7.40 Quality metrics: **a** RHD, **b** OS

Table 7.11 The settings of the good and bad points for the different examples

Example	P_{good}	P_{bad}
ZDT1/ ZDT2/ FON	$[-0.2, -0.2]$	$[1.2, 1.2]$
ZDT3	$[-0.2, -1]$	$[1.2, 1.2]$
POL	$[-2, -2]$	$[18, 28]$
QV	$[0.5, 0.5]$	$[2.3, 2.3]$
Engineering case	$[350, 0.25]$	$[750, 2.4]$

Numerical examples

Among the six numerical examples, one objective function was simply evaluated using the original function because of its simplicity, while the other objective functions were replaced by VFMs during the optimization process. As a demonstration, we use one numerical example to present a detailed comparison of the different methods. The problem formulation for the first example (ZDT2) is given as follows:

$$\begin{aligned} & minimize\, f_1(x) = x_1 \\ & \qquad \text{HF}: f_2(x) = g(x) \times h(x) \\ & \qquad \text{LF}: f_2(x) = (0.9 * g(x) + 1.1) \times (1.1 * h(x) - 0.1) \\ & where \quad g(x) = 1 + \frac{9}{n-1}\sum_{i=2}^{n} x_i \\ & \qquad h(x) = 1 - \sqrt{f_1(x)/g(x)} \\ & \qquad n = 3 \\ & \qquad 0 \le x_i \le 1, i = 1, \ldots, n \end{aligned} \tag{7.79}$$

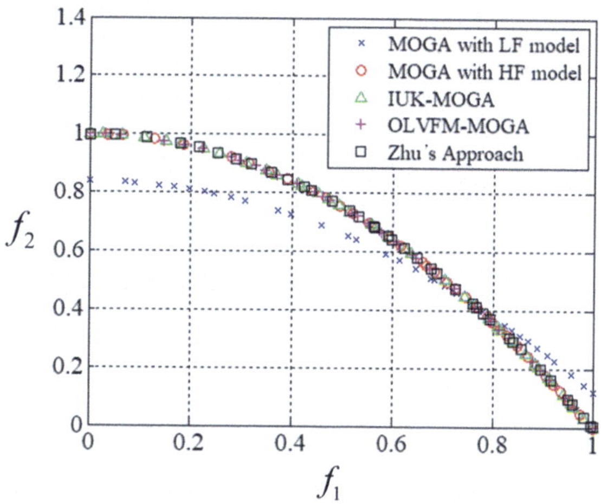

Fig. 7.41 The Pareto frontiers obtained for ZDT2 using different methods

Figure 7.41 presents a typical Pareto frontier obtained from one of the 15 runs for each of the different methods. As shown in Fig. 7.41, the Pareto frontiers obtained from the three surrogate-model-based approaches are in good agreement with the results of the MOGA with the HF model, while only a small proportion of the Pareto frontiers obtained using the MOGA with the LF model and the MOGA with HF model overlap. These findings indicate that it is difficult to obtain accurate Pareto frontiers by simply incorporating LF models directly into a MOGA.

The results for the computational efficiency of the different methods are compared in Table 7.12. In this table, the value denoted by FC represents the number of HF function calls. Note that FC is calculated by converting LF function calls into HF function calls under the assumption that the total time needed to obtain an HF sample is four times that needed to obtain an LF sample in OLVFM-MOGA.

Regarding the computational efficiency, the FC of OLVFM-MOGA is nearly 35 times lower than that of the MOGA with the HF model and 12 times lower than that of Zhu's method. Meanwhile, the average FC of OLVFM-MOGA is reduced by 25% compared to that of IUK-MOGA. The reason is that in the MOGA with the HF model, all of the individuals must be evaluated using the HF model to obtain their fitness values, while only a small portion of them must be analysed using the HF model in OLVFM-MOGA. Notably, although IUK-MOGA does not require all individuals to be analysed using the HF model during the evolution process, it still incurs a larger FC than OLVFM-MOGA does. This is because OLVFM-MOGA can make full use of the information from the LF model and reduce the surrogate model uncertainty simply by adding relatively cheap LF samples. Table 7.12 shows that Zhu's method is time-consuming when the computational cost of the LF model cannot be ignored.

Figure 7.42 depicts the numbers of newly added sample points in the first 20 generations for the MOGA with the HF model, IUK-MOGA and Zhu's method. It can be seen that the MOGA with the HF model and Zhu's method require many HF

Table 7.12 Comparison of different methods for ZDT2

Metrics	MOGA with HF model			IUK-MOGA			Zhu's approach			OLVFM-MOGA		
	30 runs	Mean	STD	30 runs	Mean	STD	30 runs	Mean	STD	30 runs	Mean	STD
RHD	[0.51 0.57]	0.55	0.01	[0.50 0.57]	0.54	0.02	[0.48 0.59]	0.55	0.02	[0.53 0.56]	0.55	0.01
OS	[0.32 0.55]	0.50	0.04	[0.26 0.68]	0.51	0.06	[0.19 0.56]	0.48	0.09	[0.41 0.54]	0.50	0.02
FC	[3785 3862]	3823	23.72	[98 210]	133	24.24	[1262.5 1292]	1277	6.81	[51.75 165.5]	104	24.83

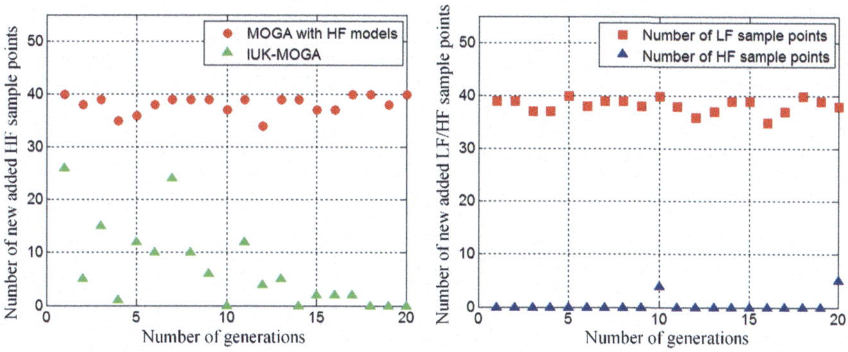

(a) The new added HF sample points in first 20 generations in MOGA with HF model and IUK-MOGA

(b) The new added LF/HF sample points in first 20 generations in Zhu's approach

Fig. 7.42 Numbers of newly added sample points in the first 20 generations for the methods considered for comparison

or LF evaluations in each generation, resulting in a cost that is unacceptable for engineering optimization problems. Figure 7.43 depicts the numbers of newly added HF/LF sample points in the first 20 generations for OLVFM-MOGA. As illustrated in Figs. 7.42 and 7.43, OLVFM-MOGA can reduce the surrogate model uncertainty by adding LF sample points, while IUK-MOGA evaluates only HF sample points. As a result, the computational cost of OLVFM-MOGA is actually less than that of IUK-MOGA.

The formulations of the remaining five numerical examples are described in Table 7.13, along with the Pareto frontiers obtained using the different methods. The results regarding the quality of the Pareto optima and the computational effort are compared in Table 7.14. As illustrated in Table 7.13, the optima obtained with IUK-MOGA, Zhu's method and OLVFM-MOGA are in good agreement with that obtained with the MOGA with the HF model. Another observation is that only a small portion of the Pareto frontiers obtained from the MOGA with the LF model and the MOGA with the HF model overlap. From Table 7.14, it can be concluded that the proposed OLVFM-MOGA method can obtain Pareto optima that are comparable to those obtained using the MOGA with the HF model. For all problems, the average computational cost of OLVFM-MOGA is much less than those of the MOGA with the HF model and Zhu's method. For problem POL, the computational efficiency of OLVFM-MOGA is slightly worse than that of IUK-MOGA. For all problems except POL, OLVFM-MOGA shows the best efficiency performance. Since OLVFM-MOGA can reduce the prediction intervals by adding LF sample points and the computational effort for solving the HF model is much larger than that for solving the LF model, the computational effort of OLVFM-MOGA is significantly less than that of the other methods in most cases.

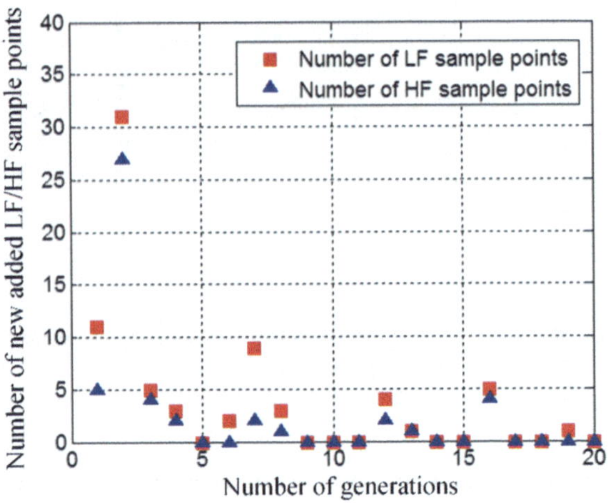

Fig. 7.43 Numbers of newly added sample points in the first 20 generations for OLVFM-MOGA

Engineering case: design optimization of a torque arm

In this analysis, a design optimization problem for a torque arm is studied. The torque arm is subjected to a bending moment and a compressive force caused by forces of P1 = 8.0 kN and P2 = 4.0 kN placed at the centre of the small end. The boundary condition is that the position of the torque arm is fixed at the hole in the large end. The Young's modulus is 200 GPa, and the Poisson's ratio is 0.3. The goal of the multi-objective optimization problem is to minimize the volume and displacement of the torque arm while keeping the stress below 190 MPa. There are six design variables, namely, α, b_1, D_1, h, t_1 and t_2, as depicted in Fig. 7.44. The other geometrical parameters remain fixed during the optimization process. Table 7.15 shows the ranges of the design variables.

During the optimization process, the calculations of the displacement and stress of the torque arm are replaced with VFM calculations. ANSYS 18.0 was selected as the simulation tool in this case. A grid consisting of approximately 5 thousand elements was selected as the LF model, and a grid consisting of approximately 50 thousand elements was chosen as the HF model, as shown in Fig. 7.45a, c. ANSYS Parametric Design Language (APDL) was used to build the geometric models and solve the finite element problems, and the simulation results are shown in Fig. 7.45b, d. The total time needed to obtain an HF sample was approximately 4 times that needed to obtain an LF sample. It can be seen from Fig. 7.45 that the simulation results of the LF and HF models vary widely under the same set of parameters; this observation indicates that the LF model cannot be used directly to obtain reliable optima.

Table 7.13 Formulations and solutions for the remaining five numerical examples

Cases	Formulation	Solutions
ZDT1	$minimize\, f_1(x) = x_1$ $\text{HF}: f_2(x) = g(x) \times h(x)$ $\text{LF}: f_2(x) = (0.8g(x) - 0.2) \times (1.2h(x) + 0.2)$ $where \quad g(x) = 1 + \frac{9}{n-1}\sum_{i=2}^{n} x_i$ $h(x) = 1 - \sqrt{f_1(x)/g(x)}$ $n = 3$ $0 \leq x_i \leq 1, i = 1, \ldots, n$	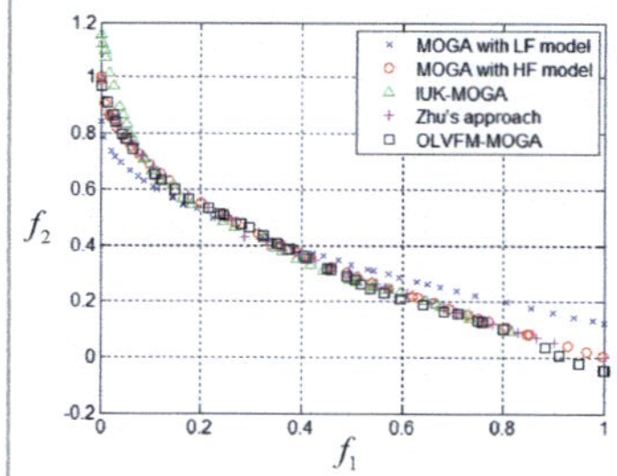
ZDT2	$minimize\, f_1(x) = x_1$ $\text{HF}: f_2(x) = g(x) \times h(x)$ $\text{LF}: f_2(x) = (0.9 * g(x) + 1.1) \times (1.1 * h(x) - 0.1)$ $where \quad g(x) = 1 + \frac{9}{n-1}\sum_{i=2}^{n} x_i$ $h(x) = 1 - \sqrt{f_1(x)/g(x)}$ $n = 3$ $0 \leq x_i \leq 1, i = 1, \ldots, n$	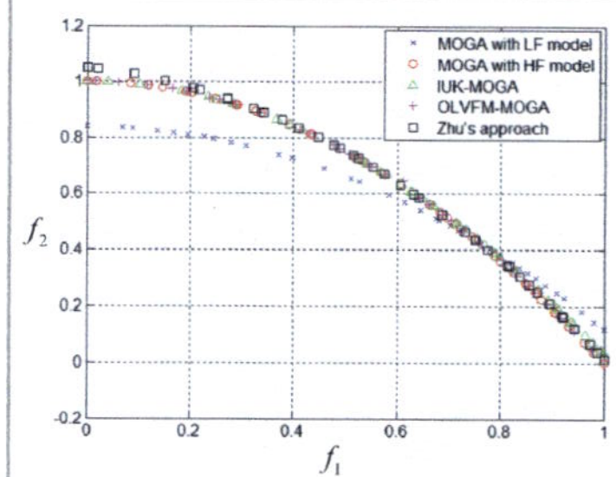

(continued)

Table 7.13 (continued)

Cases	Formulation	Solutions
ZDT3	$minimize\ f_1(x) = x_1$ $HF: f_2(x) = g(x) \times h(x)$ $LF: f_2(x) = (0.75g(x) - 0.25) \times (1.25h(x) + 0.25)$ $where\quad g(x) = 1 + \frac{9}{n-1}\sum_{i=2}^{n} x_i$ $h(x) = 1 - \sqrt{f_1(x)/g(x)} - (f_1(x)/g(x))\sin(10\pi f_1)$ $n = 3$ $0 \le x_i \le 1, i = 1, \ldots, n$	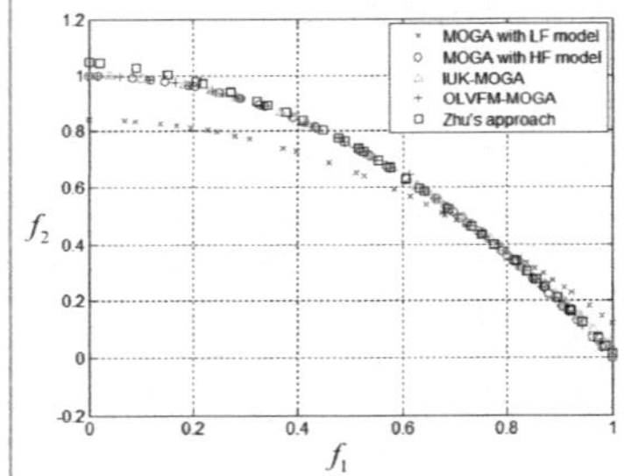
FON	$minimize\ f_1(x) = 1 - \exp(-\sum_{i=1}^{3}(x_i - \frac{1}{\sqrt{3}})^2)$ $HF: f_2(x) = 1 - \exp(-\sum_{i=1}^{3}(x_i + \frac{1}{\sqrt{3}})^2)$ $LF: f_2(x) = (1 - \exp(-(x_1 + 0.5)^2 - (x_2 + 0.55)^2$ $- (x_3 + 0.6)^2)) \cdot (1.1 + 0.25 \sin x_1)$ $-4 \le x_i \le 4, i = 1, \ldots, 3$	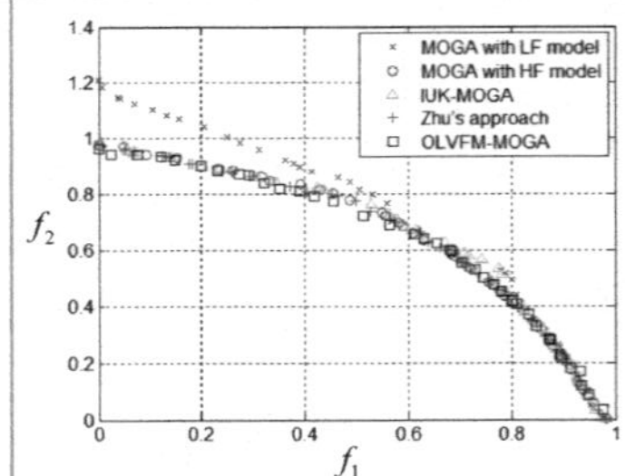

(continued)

Table 7.13 (continued)

Cases	Formulation	Solutions
POL	*minimize* HF : $f_1(x) = [1 + (A_1 - B_1)^2 + (A_1 - B_1)^2]$ LF : $f_1(x) = [1 + (0.9A_1 - 1.2B_1)^2 + 0.9(1.2A_2 - 0.9B_2)^2]$ $f_2(x) = (x_1 + 3)^2 + (x_2 + 1)^2$ *where* $A_1 = 0.5\sin 1 - 2\cos 1 + \sin 2 - 1.5\cos 2$ $A_2 = 1.5\sin 1 - \cos 1 + 2\sin 2 - 0.5\cos 2$ $B_1 = 0.5\sin x_1 - 2\cos x_1 + \sin x_2 - 1.5\cos x_2$ $B_2 = 0.5\sin x_1 - 2\cos x_1 + \sin x_2 - 1.5\cos x_2$ $-\pi \le x_i \le \pi, i = 1, 2$	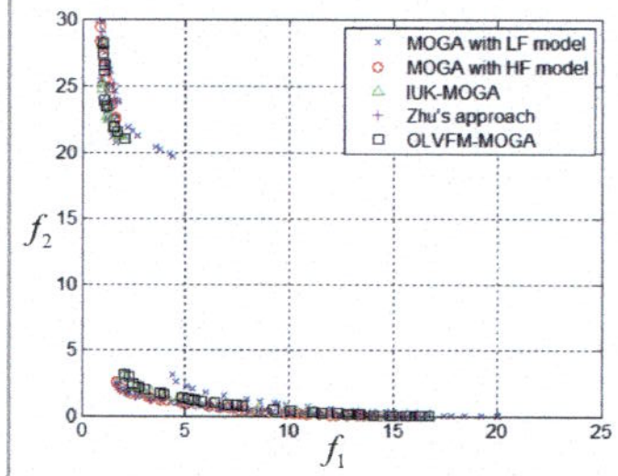
QV	*minimize* $f_1(x) = (1/n\sum_{i=1}^{n}(x_i^2 - 20\pi x_i + 10))^{1/4}$ HF : $f_2(x) = (1/n\sum_{i=1}^{n}((x_i - 1.5)^2 - 20\pi(x_i - 1.5) + 10))^{1/4}$ LF : $f_2(x) = (1/n((0.9(x_1 - 1.5)^2 - 20\pi(x_1 - 1.5) + 10)$ $+ (1.1(x_2 - 1.5)^2 - 20\pi(x_2 - 1.5) + 10)$ $+ (0.9(x_3 - 1.5)^2 - 20\pi(x_3 - 1.5) + 10)$ $+ (1.1(x_4 - 1.5)^2 - 20\pi(x_4 - 1.5) + 10)$ $+ (0.9(x_5 - 1.5)^2 - 20\pi(x_5 - 1.5) + 10)))^{1/4}$ *where* $n = 5$ $-5 \le x_i \le 5, i = 1, \ldots, n$	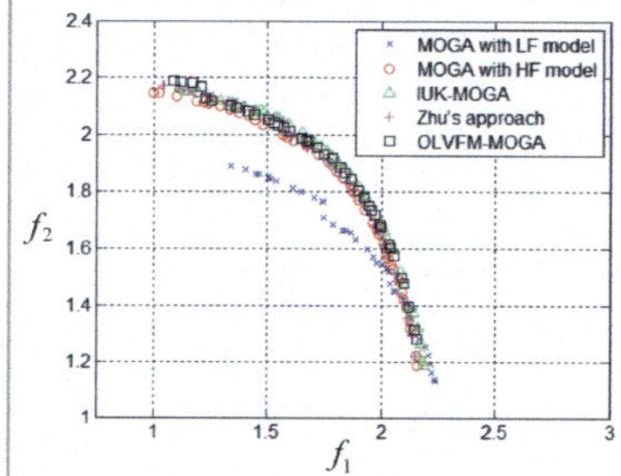

Table 7.14 Comparison of the different methods on the remaining five numerical examples

	Metrics	MOGA with HF model			IUK-MOGA			Zhu's approach			OLVFM-MOGA		
		30 runs	Mean	STD	30 runs	Mean	STD	30 runs	Mean	STD	30 runs	Mean	STD
ZDT1	RHD	[0.34 0.39]	0.38	0.01	[0.32 0.45]	0.37	0.03	[0.34 0.41]	0.39	0.01	[0.35 0.40]	0.37	0.02
	OS	[0.35 0.51]	0.50	0.03	[0.22 0.67]	0.44	0.11	[0.32 0.59]	0.51	0.05	[0.29 0.57]	0.46	0.07
	FC	[3823 3864]	3840.93	10.46	[191 476]	323.63	81.55	[1284.5 1336.5]	1304.17	12.61	[137 439.5]	257.96	58.55
ZDT2	RHD	[0.51 0.57]	0.55	0.01	[0.50 0.57]	0.54	0.02	[0.48 0.59]	0.55	0.02	[0.53 0.56]	0.55	0.01
	OS	[0.32 0.55]	0.50	0.04	[0.26 0.68]	0.51	0.06	[0.19 0.56]	0.48	0.09	[0.41 0.54]	0.50	0.02
	FC	[3785 3862]	3823	23.72	[98 210]	133	24.24	[1262.5 1292]	1277	6.81	[51.75 165.5]	104	24.83
ZDT3	RHD	[0.28 0.45]	0.37	0.06	[0.26 0.57]	0.40	0.07	[0.34 0.48]	0.41	0.04	[0.27 0.55]	0.39	0.06
	OS	[0.17 0.52]	0.38	0.15	[0.13 0.50]	0.33	0.11	[0.14 0.52]	0.33	0.13	[0.16 0.49]	0.34	0.09
	FC	[3817 3874]	3840.23	15.02	[141 473]	312.10	82.40	[1268.75 1304.5]	1283.13	10.25	[142 453.75]	293.73	67.16
FON	RHD	[0.54 0.59]	0.56	0.01	[0.51 0.58]	0.56	0.01	[0.53 0.56]	0.55	0.01	[0.54 0.58]	0.56	0.01
	OS	[0.31 0.49]	0.46	0.04	[0.15 0.49]	0.42	0.10	[0.28 0.49]	0.46	0.06	[0.29 0.49]	0.42	0.07
	FC	[3819 3876]	3844.93	11.96	[137 348]	249.20	54.48	[1331 1362.8]	1347.77	7.27	[83.75 292]	169.91	55.48

(continued)

Table 7.14 (continued)

	Metrics	MOGA with HF model			IUK-MOGA			Zhu's approach			OLVFM-MOGA		
		30 runs	Mean	STD	30 runs	Mean	STD	30 runs	Mean	STD	30 runs	Mean	STD
POL	RHD	[0.22 0.24]	0.23	0.01	[0.23 0.26]	0.24	0.01	[0.23 0.27]	0.25	0.01	[0.23 0.26]	0.24	0.01
	OS	[0.61 0.71]	0.63	0.04	[0.65 0.74]	0.67	0.03	[0.58 0.75]	0.64	0.04	[0.61 0.74]	0.67	0.04
	FC	[3813 3870]	3841.33	14.54	[78 142]	101.30	17.03	[1273.25 1292.5]	1283.36	3.98	[83.75 130.25]	104.90	11.61
QV	RHD	[0.73 0.80]	0.76	0.02	[0.74 0.88]	0.79	0.03	[0.72 0.83]	0.77	0.02	[0.71 0.87]	0.79	0.04
	OS	[0.23 0.41]	0.33	0.05	[0.10 0.36]	0.23	0.06	[0.19 0.40]	0.28	0.05	[0.11 0.39]	0.23	0.07
	FC	[5693 5760]	5738.80	13.85	[114 548]	249.13	89.73	[1809.75 1938.5]	1887.12	32.53	[108 437.25]	235.99	81.91

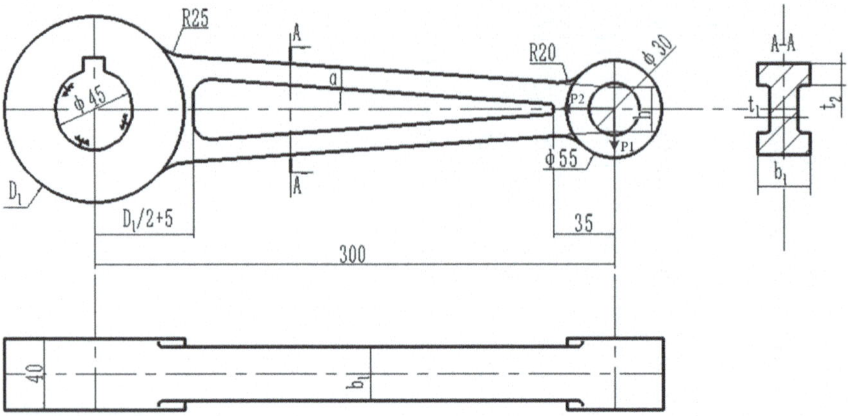

Fig. 7.44 Geometry and parameterization of the torque arm

Table 7.15 Ranges of the design variables

Design variables	Lower bound (mm)	Upper bound (mm)
α	3.0	4.5
b_1	25.0	35.0
D_1	90.0	120.0
h	20.0	30.0
t_1	12.0	22.0
t_2	8.0	12.0

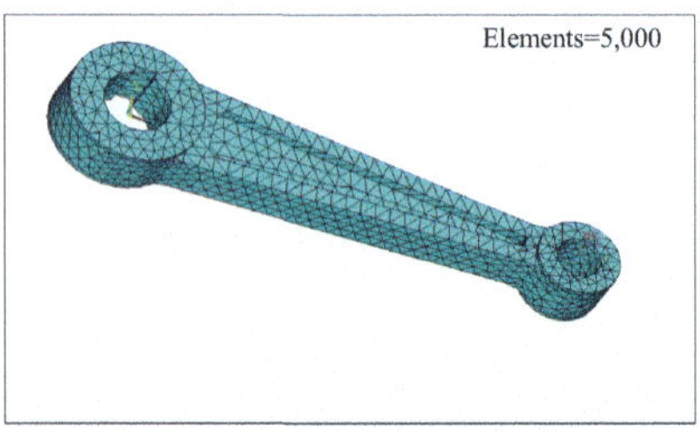

(a) LF simulation model

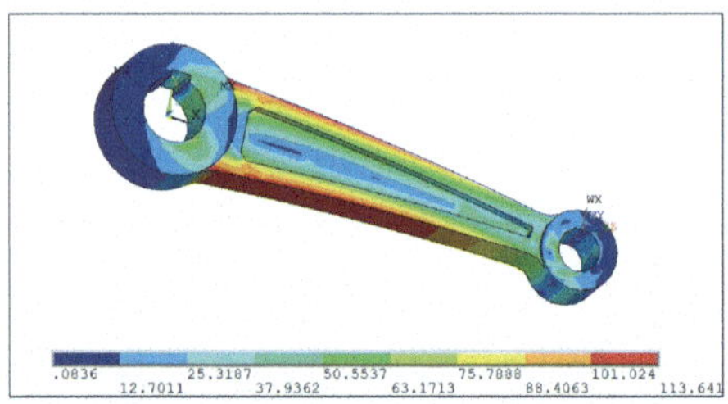

(b) Simulation result of LF simulation model

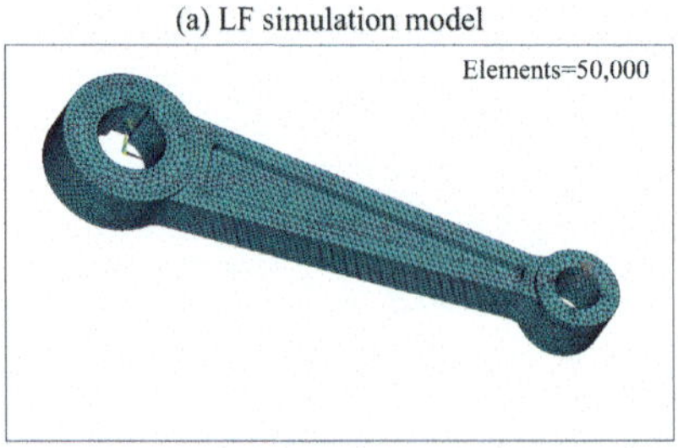

(c) HF simulation model

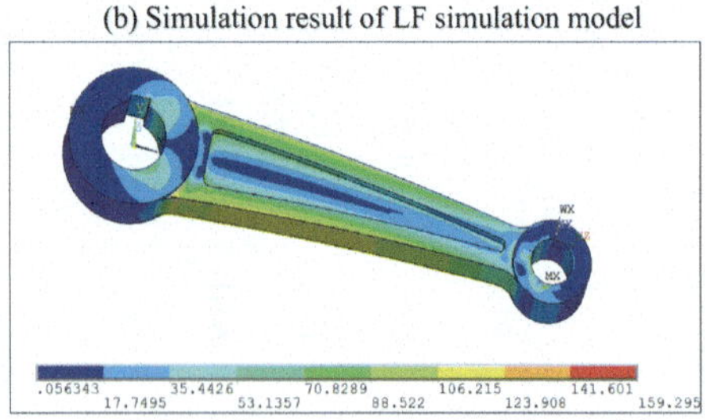

(d) Simulation result of HF simulation model

Fig. 7.45 Simulation results obtained with the LF and HF models

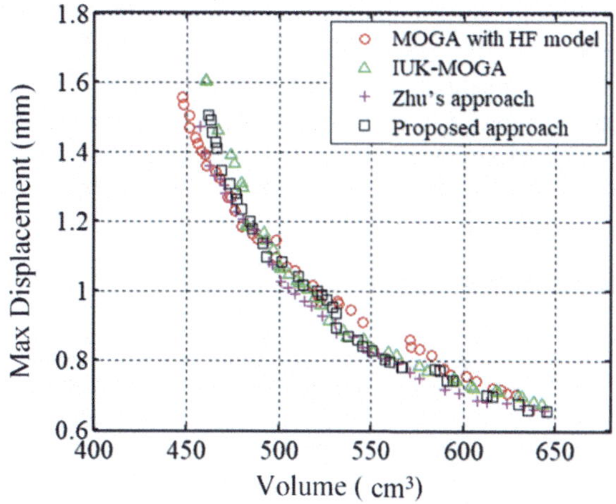

Fig. 7.46 Pareto frontiers obtained using different methods

Fifteen runs were performed with each method to account for the influence of randomness in IUK-MOGA and OLVFM-MOGA. Figure 7.46 illustrates typical Pareto frontiers obtained using these methods.

As shown in Fig. 7.46, the Pareto frontiers obtained with the two surrogate-model-based approaches are in good agreement with that obtained using the MOGA with the HF model. To further demonstrate the superiority of the proposed approach, the convergence quality and diversity of the Pareto optima and the levels of computational efforts required for the different methods were assessed, and the results are summarized in Table 7.15.

As illustrated in Table 7.16, the average values of the RHD and OS for the other three approaches are close to those for the MOGA with the HF model, indicating that these three approaches can obtain Pareto optima with convergence and diversity comparable to those of the Pareto optima obtained with the MOGA with the HF model.

Table 7.16 Comparison of the results of the different methods

	MOGA with HF model	IUK-MOGA			Zhu's approach			OLVFM-MOGA		
		15 runs	Mean	STD	15 runs	Mean	STD	15 runs	Mean	STD
RHD	0.37	[0.31 0.40]	0.35	0.04	[0.28 0.31]	0.29	0.005	[0.33 0.43]	0.36	0.03
OS	0.22	[0.14 0.32]	0.23	0.07	[0.14 0.21]	0.17	0.015	[0.21 0.40]	0.25	0.06
FC	3840	[127 429]	174.73	91.37	[1386 1395]	1391	3.137	[100 216.25]	143.22	32.01

Regarding the computational efficiency, the FC of OLVFM-MOGA is nearly 25 times lower than that of the MOGA with the HF model and 10 times lower than that of Zhu's method. Meanwhile, the average FC of OLVFM-MOGA is reduced by 25% compared to that of IUK-MOGA. Moreover, IUK-MOGA shows worse stability performance.

References

Akhtar T, Shoemaker CA (2016) Multi objective optimization of computationally expensive multi-modal functions with RBF surrogates and multi-rule selection. J Global Optim 64:17–32

Alexandrov NM, Dennis J, Lewis RM, Torczon V (1998) A trust-region framework for managing the use of approximation models in optimization. Struct Optim 15:16–23

Andrés E, Salcedo-Sanz S, Monge F, Pérez-Bellido AM (2012) Efficient aerodynamic design through evolutionary programming and support vector regression algorithms. Expert Syst Appl 39:10700–10708

Apley DW, Liu J, Chen W (2006) Understanding the effects of model uncertainty in robust design with computer experiments. J Mech Des 128:945–958

Arendt PD, Apley DW, Chen W (2013) Objective-oriented sequential sampling for simulation based robust design considering multiple sources of uncertainty. J Mech Des 135:051005

Audet C, Denni J, Moore D, Booker A, Frank P (2000) A surrogate-model-based method for constrained optimization. In: 8th symposium on multidisciplinary analysis and optimization, p 4891

Bahrami S, Tribes C, Devals C, Vu T, Guibault F (2016) Multi-fidelity shape optimization of hydraulic turbine runner blades using a multi-objective mesh adaptive direct search algorithm. Appl Math Model 40:1650–1668

Basudhar A, Dribusch C, Lacaze S, Missoum S (2012) Constrained efficient global optimization with support vector machines. Struct Multidiscip Optim 46:201–221

Branke J, Schmidt C (2005) Faster convergence by means of fitness estimation. Soft Comput 9:13–20

Bui LT, Abbass HA, Essam D (2005) Fitness inheritance for noisy evolutionary multi-objective optimization. In: Proceedings of the 7th annual conference on Genetic and evolutionary computation: ACM, pp 779–785

Chaudhuri A, Haftka RT (2014) Efficient global optimization with adaptive target setting. AIAA J 52:1573–1578

Chen T-Y, Cheng Y-L (2010) Data-mining assisted structural optimization using the evolutionary algorithm and neural network. Eng Optim 42:205–222

Chen J-H, Goldberg DE, Ho S-Y, Sastry K (2002) Fitness inheritance in multi-objective optimization. In: GECCO, pp 319–326

Chen W, Jin R, Sudjianto A (2005) Analytical variance-based global sensitivity analysis in simulation-based design under uncertainty. J Mech Des 127:875–886

Chen G, Han X, Liu G, Jiang C, Zhao Z (2012) An efficient multi-objective optimization method for black-box functions using sequential approximate technique. Appl Soft Comput 12:14–27

Chen Z, Qiu H, Gao L, Li X, Li P (2014) A local adaptive sampling method for reliability-based design optimization using Kriging model. Struct Multidiscip Optim 49:401–416

Chen S, Jiang Z, Yang S, Apley DW, Chen W (2016) Nonhierarchical multi-model fusion using spatial random processes. Int J Numer Meth Eng 106:503–526

Cheng J, Liu Z, Wu Z, Li X, Tan J (2015a) Robust optimization of structural dynamic characteristics based on adaptive Kriging model and CNSGA. Struct Multidiscip Optim 51:423–437

Cheng R, Jin Y, Narukawa K, Sendhoff B (2015b) A multiobjective evolutionary algorithm using gaussian process-based inverse modeling. IEEE Trans Evol Comput 19:838–856

Cheng S, Zhou J, Li M (2015c) A new hybrid algorithm for multi-objective robust optimization with interval uncertainty. J Mech Des 137:021401

Chung H-S, Alonso JJ (2004) Multiobjective optimization using approximation model-based genetic algorithms. AIAA paper 4325

Clarke SM, Griebsch JH, Simpson TW (2005) Analysis of support vector regression for approximation of complex engineering analyses. J Mech Des 127:1077–1087

Coello CAC (2000) Use of a self-adaptive penalty approach for engineering optimization problems. Comput Ind 41:113–127

Couckuyt I, Deschrijver D, Dhaene T (2014) Fast calculation of multiobjective probability of improvement and expected improvement criteria for Pareto optimization. J Glob Optim 60:575–594

Datta R, Regis RG (2016) A surrogate-assisted evolution strategy for constrained multi-objective optimization. Expert Syst Appl 57:270–284

Deb K, Pratap A, Agarwal S, Meyarivan T (2002) A fast and elitist multiobjective genetic algorithm: NSGA-II. IEEE Trans Evol Comput 6:182–197

Desautels T, Krause A, Burdick JW (2014) Parallelizing exploration-exploitation tradeoffs in gaussian process bandit optimization. J Mach Learn Res 15:3873–3923

Du X, Chen W (2004) Sequential optimization and reliability assessment method for efficient probabilistic design. J Mech Des 126:225–233

Du X, Guo J, Beeram H (2008) Sequential optimization and reliability assessment for multidisciplinary systems design. Struct Multidiscip Optim 35:117–130

Eldred M, Giunta A, Wojtkiewicz S, Trucano T (2002) Formulations for surrogate-based optimization under uncertainty. In: 9th AIAA/ISSMO symposium on multidisciplinary analysis and optimization, p 5585

Forrester AIJ, Keane AJ, Bressloff NW (2006) Design and analysis of "Noisy" computer experiments. AIAA J 44:2331–2339

Forrester AIJ, Keane AJ (2009) Recent advances in surrogate-based optimization. Prog Aerosp Sci 45:50–79

Forrester A, Sobester A, Keane A (2008) Engineering design via surrogate modelling: a practical guide. Wiley

Gano SE, Renaud JE, Agarwal H, Tovar A (2006a) Reliability-based design using variable-fidelity optimization. Struct Infrastruct Eng 2:247–260

Gano SE, Renaud JE, Martin JD, Simpson TW (2006b) Update strategies for kriging models used in variable fidelity optimization. Struct Multidiscip Optim 32:287–298

Gary Wang G, Dong Z, Aitchison P (2001) Adaptive response surface method-a global optimization scheme for approximation-based design problems. Eng Optim 33:707–733

Ghisu T, Parks GT, Jarrett JP, Clarkson PJ (2011) Robust design optimization of gas turbine compression systems. J Propul Power 27:282–295

Gluzman S, Yukalov V (2006) Self-similar power transforms in extrapolation problems. J Math Chem 39:47–56

Goel T, Vaidyanathan R, Haftka RT, Shyy W, Queipo NV, Tucker K (2007) Response surface approximation of Pareto optimal front in multi-objective optimization. Comput Methods Appl Mech Eng 196:879–893

Gu L (2001) A comparison of polynomial based regression models in vehicle safety analysis. In: ASME design engineering technical conferences, ASME paper no.: DETC/DAC-21083

Gu X, Renaud JE, Batill SM, Brach RM, Budhiraja AS (2000) Worst case propagated uncertainty of multidisciplinary systems in robust design optimization. Struct Multidiscip Optim 20:190–213

Gunawan S (2004) Parameter sensitivity measures for single objective, multi-objective, and feasibility robust design optimization

Gunawan S, Azarm S (2004) Non-gradient based parameter sensitivity estimation for single objective robust design optimization. J Mech Des 126:395–402

Gunawan S, Azarm S (2005a) A feasibility robust optimization method using sensitivity region concept. J Mech Des 127:858–865
Gunawan S, Azarm S (2005b) Multi-objective robust optimization using a sensitivity region concept. Struct Multidiscip Optim 29:50–60
Gutmann HM (2001) A radial basis function method for global optimization. J Global Optim 19:201–227
Haftka RT (1991) Combining global and local approximations. AIAA J 29:1523–1525
Hamdaoui M, Oujebbour F-Z, Habbal A, Breitkopf P, Villon P (2015) Kriging surrogates for evolutionary multi-objective optimization of CPU intensive sheet metal forming applications. IntJ Mater Form 8:469–480
Han Z-H, Zimmermann R, Goretz S (2010) A new cokriging method for variable-fidelity surrogate modeling of aerodynamic data
Han Z, Zimmerman R, Görtz S (2012) Alternative cokriging method for variable-fidelity surrogate modeling. AIAA J 50:1205–1210
Hsu YL, Wang SG, Yu CC (2003) A sequential approximation method using neural networks for engineering design optimization problems. Eng Optim 35:489–511
Hu Z, Mahadevan S (2017) Uncertainty quantification in prediction of material properties during additive manufacturing. Scripta Mater 135:135–140
Hu W, Enying L, Yao LG (2008) Optimization of drawbead design in sheet metal forming based on intelligent sampling by using response surface methodology. J Mater Process Technol 206:45–55
Hu W, Li M, Azarm S, Almansoori A (2011) Multi-objective robust optimization under interval uncertainty using online approximation and constraint cuts. J Mech Des 133:061002
Huang D, Allen TT, Notz WI, Miller RA (2006a) Sequential kriging optimization using multiple-fidelity evaluations. Struct Multidiscip Optim 32:369–382
Huang D, Allen TT, Notz WI, Zeng N (2006b) Global optimization of stochastic black-box systems via sequential kriging meta-models. J Global Optim 34:441–466
Jeong S, Minemura Y, Obayashi S (2006) Optimization of combustion chamber for diesel engine using kriging model. J Fluid Sci Technol 1:138–146
Jin Y (2005) A comprehensive survey of fitness approximation in evolutionary computation. Soft Comput 9:3–12
Jin Y (2011) Surrogate-assisted evolutionary computation: recent advances and future challenges. Swarm Evol Comput 1:61–70
Jin Y, Olhofer M, Sendhoff B (2000) On evolutionary optimization with approximate fitness functions. In: GECCO, pp 786–793
Jin Y, Olhofer M, Sendhoff B (2002) A framework for evolutionary optimization with approximate fitness functions. IEEE Trans Evol Comput 6:481–494
Jones DR (2001) A taxonomy of global optimization methods based on response surfaces. J Glob Optim 21:345–383
Jones DR, Schonlau M, Welch WJ (1998) Efficient global optimization of expensive black-box functions. J Glob Optim 13:455–492
Kalyanmoy D (2001) Multi objective optimization using evolutionary algorithms: John Wiley and Sons
Kannan B, Kramer SN (1994) An augmented Lagrange multiplier based method for mixed integer discrete continuous optimization and its applications to mechanical design. J Mech Des 116:405–411
Keane AJ (2006) Statistical improvement criteria for use in multiobjective design optimization. AIAA J 44:879–891
Kerschen G, Worden K, Vakakis AF, Golinval J-C (2006) Past, present and future of nonlinear system identification in structural dynamics. Mech Syst Signal Process 20:505–592
Kim N-K, Kim D-H, Kim D-W, Kim H-G, Lowther D, Sykulski JK (2010) Robust optimization utilizing the second-order design sensitivity information. IEEE Trans Magn 46:3117–3120
Kitayama S, Srirat J, Arakawa M, Yamazaki K (2013) Sequential approximate multi-objective optimization using radial basis function network. Struct Multidiscip Optim 48:501–515

Kleijnen JP, Van Beers W, Van Nieuwenhuyse I (2012) Expected improvement in efficient global optimization through bootstrapped kriging. J Glob Optim 54:59–73

Koch P, Yang R-J, Gu L (2004) Design for six sigma through robust optimization. Struct Multidiscip Optim 26:235–248

Kodiyalam S, Nagendra S, DeStefano J (1996) Composite sandwich structure optimization with application to satellite components. AIAA J 34:614–621

Laurenceau J, Meaux M, Montagnac M, Sagaut P (2010) Comparison of gradient-based and gradient-enhanced response-surface-based optimizers. AIAA J 48:981–994

Le MN, Ong YS, Menzel S, Jin Y, Sendhoff B (2013) Evolution by adapting surrogates. Evol Comput 21:313–340

Lee K-H, Park G-J (2001) Robust optimization considering tolerances of design variables. Comput Struct 79:77–86

Li G (2007) Online and offline approximations for population based multi-objective optimization. ProQuest

Li M (2011) An improved kriging-assisted multi-objective genetic algorithm. J Mech Des 133:071008-071008-071011

Li M, Azarm S, Boyars A (2006) A new deterministic approach using sensitivity region measures for multi-objective robust and feasibility robust design optimization. J Mech Des 128:874–883

Li G, Li M, Azarm S, Rambo J, Joshi Y (2007) Optimizing thermal design of data center cabinets with a new multi-objective genetic algorithm. Distrib Parallel Databases 21:167–192

Li G, Li M, Azarm S, Al Hashimi S, Al Ameri T, Al Qasas N (2009a) Improving multi-objective genetic algorithms with adaptive design of experiments and online metamodeling. Struct Multidiscip Optim 37:447–461

Li M, Williams N, Azarm S (2009b) Interval uncertainty reduction and single-disciplinary sensitivity analysis with multi-objective optimization. J Mech Des 131:031007

Li M, Hamel J, Azarm S (2010) Optimal uncertainty reduction for multi-disciplinary multi-output systems using sensitivity analysis. Struct Multidiscip Optim 40:77–96

Li X, Qiu H, Chen Z, Gao L, Shao X (2016) A local Kriging approximation method using MPP for reliability-based design optimization. Comput Struct 162:102–115

Lian Y, Liou M-S (2005) Multiobjective optimization using coupled response surface model and evolutionary algorithm. AIAA J 43:1316–1325

Lim J, Lee B, Lee I (2014) Second-order reliability method-based inverse reliability analysis using Hessian update for accurate and efficient reliability-based design optimization. Int J Numer Meth Eng 100:773–792

Liu Y, Collette M (2014) Improving surrogate-assisted variable fidelity multi-objective optimization using a clustering algorithm. Appl Soft Comput 24:482–493

Liu J, Han Z, Song W (2012) Comparison of infill sampling criteria in kriging-based aerodynamic optimization. In: 28th congress of the international council of the aeronautical sciences, pp 23–28

Long T, Wu D, Guo XS, Wang GG, Liu L (2015) Efficient adaptive response surface method using intelligent space exploration strategy. Struct Multidiscip Optim 51:1335–1362

Madsen HO, Krenk S, Lind NC (2006) Methods of structural safety. Courier Corporation

Martin JD, Simpson TW (2005) Use of kriging models to approximate deterministic computer models. AIAA J 43:853–863

McKay MD, Beckman RJ, Conover WJ (2000) A comparison of three methods for selecting values of input variables in the analysis of output from a computer code. Technometrics 42:55–61

Mlakar M, Petelin D, Tušar T, Filipič B (2015) GP-DEMO: differential evolution for multiobjective optimization based on Gaussian process models. Eur J Oper Res 243:347–361

Mogilicharla A, Mittal P, Majumdar S, Mitra K (2015) Kriging surrogate based multi-objective optimization of bulk vinyl acetate polymerization with branching. Mater Manuf Processes 30:394–402

Oberkampf W, Helton J, Sentz K (2001) Mathematical representation of uncertainty. In: 19th AIAA applied aerodynamics conference, p 1645

Papadimitriou D, Giannakoglou K (2013) Third-order sensitivity analysis for robust aerodynamic design using continuous adjoint. Int J Numer Meth Fluids 71:652–670
Park H-S, Dang X-P (2010) Structural optimization based on CAD–CAE integration and metamodeling techniques. Comput Aided Des 42:889–902
Parr J, Keane A, Forrester AI, Holden C (2012) Infill sampling criteria for surrogate-based optimization with constraint handling. Eng Optim 44:1147–1166
Ponweiser W, Wagner T, Vincze M (2008) Clustered multiple generalized expected improvement: a novel infill sampling criterion for surrogate models. In: IEEE world congress on computational intelligence evolutionary computation, 2008. CEC 2008, pp 3515–3522. IEEE
Ratle A (1998) Accelerating the convergence of evolutionary algorithms by fitness landscape approximation. Parallel problem solving from nature—PPSN V. Springer, pp 87–96
Regis RG (2013) Evolutionary programming for high-dimensional constrained expensive black-box optimization using radial basis functions. IEEE Trans Evol Comput 18:326–347
Regis RG, Shoemaker CA (2005) Constrained global optimization of expensive black box functions using radial basis functions. J Glob Optim 31:153–171
Regis RG, Shoemaker CA (2007) Improved strategies for radial basis function methods for global optimization. J Glob Optim 37:113–135
Regis RG, Shoemaker CA (2013) Combining radial basis function surrogates and dynamic coordinate search in high-dimensional expensive black-box optimization. Eng Optim 45:529–555
Renaud J (1997) Automatic differentiation in robust optimization. AIAA J 35:1072–1079
Sacks J, Welch WJ, Mitchell TJ, Wynn HP (1989) Design and analysis of computer experiments. Stat Sci 409–423
Sasena MJ (2002) Flexibility and efficiency enhancements for constrained global design optimization with kriging approximations. Citeseer
Sasena MJ, Papalambros P, Goovaerts P (2002) Exploration of metamodeling sampling criteria for constrained global optimization. Eng Optim 34:263–278
Schonlau M (1997) Computer experiments and global optimization
Shan S, Wang GG (2005) An efficient Pareto set identification approach for multiobjective optimization on black-box functions. J Mech Des 127:866–874
Sharif B, Wang GG, ElMekkawy TY (2008) Mode pursuing sampling method for discrete variable optimization on expensive black-box functions. J Mech Des 130:021402
Shimoyama K, Sato K, Jeong S, Obayashi S (2013) Updating kriging surrogate models based on the hypervolume indicator in multi-objective optimization. J Mech Des 135:094503
Shu L, Jiang P, Zhou Q, Shao X, Hu J, Meng X (2018) An on-line variable fidelity metamodel assisted multi-objective genetic algorithm for engineering design optimization. Appl Soft Comput 66:438–448
Simpson TW, Mauery TM, Korte JJ, Mistree F (2001) Kriging models for global approximation in simulation-based multidisciplinary design optimization. AIAA J 39:2233–2241
Simpson TW, Booker AJ, Ghosh D, Giunta AA, Koch PN, Yang RJ (2004) Approximation methods in multidisciplinary analysis and optimization: a panel discussion. Struct Multidiscip Optim 27:302–313
Sóbester A, Leary SJ, Keane AJ (2005) On the design of optimization strategies based on global response surface approximation models. J Global Optim 33:31–59
Song Z, Murray BT, Sammakia B, Lu S (2012) Multi-objective optimization of temperature distributions using Artificial Neural Networks. In: 2012 13th IEEE intersociety conference on thermal and thermomechanical phenomena in electronic systems (ITherm), pp 1209–1218. IEEE
Srinivas N, Krause A, Kakade SM, Seeger MW (2012) Information-theoretic regret bounds for gaussian process optimization in the bandit setting. IEEE Trans Inf Theory 58:3250–3265
Steuben JC, Turner CJ (2015) Graph analysis of non-uniform rational B-spline-based metamodels. Eng Optim 47:1157–1176
Sun GY, Li GY, Gong ZH, He GQ, Li Q (2011) Radial basis functional model for multi-objective sheet metal forming optimization. Eng Optim 43:1351–1366

Sun X, Gong D, Jin Y, Chen S (2013) A new surrogate-assisted interactive genetic algorithm with weighted semisupervised learning. IEEE Trans Cybern 43:685–698

Sun C, Jin Y, Cheng R, Ding J, Zeng J (2017) Surrogate-assisted cooperative swarm optimization of high-dimensional expensive problems. IEEE Trans Evol Comput

Taguchi G (1978) Performance analysis design. Int J Prod Res 16:521–530

Tan MHY (2015a) Robust parameter design with computer experiments using orthonormal polynomials. Technometrics 57:468–478

Tan MHY (2015b) Stochastic polynomial interpolation for uncertainty quantification with computer experiments. Technometrics 57:457–467

Tang Y, Chen J, Wei J (2013) A surrogate-based particle swarm optimization algorithm for solving optimization problems with expensive black box functions. Eng Optim 45:557–576

Törn A, Zilinskas A (1989) Global optimization. Springer

Viana F, Haftka R (2010) Surrogate-based optimization with parallel simulations using the probability of improvement. In: 13th AIAA/ISSMO multidisciplinary analysis optimization conference, p 9392

Viana FA, Simpson TW, Balabanov V, Toropov V (2014) Special section on multidisciplinary design optimization: metamodeling in multidisciplinary design optimization: how far have we really come? AIAA J 52:670–690

Wang GG (2003) Adaptive response surface method using inherited latin hypercube design points. J Mech Des 125:210–220

Wang GG, Shan S (2007) Review of metamodeling techniques in support of engineering design optimization. J Mech Des 129:370–380

Wang L, Shan S, Wang GG (2004) Mode-pursuing sampling method for global optimization on expensive black-box functions. Eng Optim 36:419–438

Wang H, Jin Y, Jansen JO (2016) Data-driven surrogate-assisted multiobjective evolutionary optimization of a trauma system. IEEE Trans Evol Comput 20:939–952

Wang H, Jin Y, Doherty J (2017) Committee-based active learning for surrogate-assisted particle swarm optimization of expensive problems. IEEE Trans Cybern

Wu J, Azarm S (2001) Metrics for quality assessment of a multiobjective design optimization solution set. J Mech Des 123:18–25

Xia T, Li M, Zhou J (2016) A sequential robust optimization approach for multidisciplinary design optimization with uncertainty. J Mech Des 138:111406

Xiao S, Rotaru M, Sykulski JK (2012) Exploration versus exploitation using kriging surrogate modelling in electromagnetic design. COMPEL Int J Comput Math Electr Electron Eng 31:1541–1551

Xiao S, Rotaru M, Sykulski JK (2013) Adaptive weighted expected improvement with rewards approach in kriging assisted electromagnetic design. IEEE Trans Magn 49:2057–2060

Yondo R, Andrés E, Valero E (2018) A review on design of experiments and surrogate models in aircraft real-time and many-query aerodynamic analyses. Prog Aerosp Sci 96:23–61

Younis A, Dong Z (2010) Trends, features, and tests of common and recently introduced global optimization methods. Eng Optim 42:691–718

Zhang S, Zhu P, Chen W, Arendt P (2012) Concurrent treatment of parametric uncertainty and metamodeling uncertainty in robust design. Struct Multidiscip Optim 47:63–76

Zheng J, Li Z, Gao L, Jiang G (2016) A parameterized lower confidence bounding scheme for adaptive metamodel-based design optimization. Eng Comput 33:2165–2184

Zhou J, Li M (2014) Advanced robust optimization with interval uncertainty using a single-looped structure and sequential quadratic programming. J Mech Des 136:021008

Zhou J, Cheng S, Li M (2012) Sequential quadratic programming for robust optimization with interval uncertainty. J Mech Des 134:100913

Zhou Q, Shao X, Jiang P, Cao L, Zhou H, Shu L (2015a) Differing mapping using ensemble of metamodels for global variable-fidelity metamodeling. CMES Comput Model Eng Sci 106: 23–355

Zhou Q, Shao X, Jiang P, Zhou H, Cao L, Zhang L (2015b) A deterministic robust optimisation method under interval uncertainty based on the reverse model. J Eng Des 26:416–444

Zhou Q, Shao X, Jiang P, Zhou H, Cao L, Zhang L (2015) A deterministic robust optimisation method under interval uncertainty based on the reverse model. J Eng Des 1–29

Zhou Q, Shao X, Jiang P, Gao Z, Wang C, Shu L (2016a) An active learning metamodeling approach by sequentially exploiting difference information from variable-fidelity models. Adv Eng Inform 30:283–297

Zhou Q, Shao X, Jiang P, Gao Z, Zhou H, Shu L (2016b) An active learning variable-fidelity metamodelling approach based on ensemble of metamodels and objective-oriented sequential sampling. J Eng Des 27:205–231

Zhou Q, Wang Y, Jiang P, Shao X, Choi S-K, Hu J, Cao L, Meng X (2017) An active learning radial basis function modeling method based on self-organization maps for simulation-based design problems. Knowl Based Syst 131:10–27

Zhu P, Zhang Y, Chen G (2009) Metamodel-based lightweight design of an automotive front-body structure using robust optimization. Proc Inst Mech Eng Part D J Automob Eng 223:1133–1147

Zhu J, Wang Y-J, Collette M (2013) A multi-objective variable-fidelity optimization method for genetic algorithms. Eng Optim 46:521–542

Zhu J, Wang Y-J, Collette M (2014) A multi-objective variable-fidelity optimization method for genetic algorithms. Eng Optim 46:521–542

Zhu P, Zhang S, Chen W (2015) Multi-point objective-oriented sequential sampling strategy for constrained robust design. Eng Optim 47:287–307

Zimmermann R, Han Z (2010) Simplified cross-correlation estimation for multi-fidelity surrogate cokriging models. Adv Appl Math Sci 7:181–202

Chapter 8
Conclusion

This book is focused on surrogate-model-based engineering design and optimization. The intent of writing this book is to provide a systematic knowledge in the area of surrogate modelling methods and surrogate-model-based optimization algorithms for the design of engineering products that require computationally expensive simulations. The hope is that it will promote the development and enrichment of surrogate model approaches to accelerate the design process and reduce simulation costs. The first seven chapters give an overall introduction to the concept of surrogate models, different types of surrogate models, some surrogate-model-related issues and the applications of surrogate models in design and optimization. The contents of each chapter can be summarized as follows.

Chapter 1 started from the perspective of the practice of mechanical product design and showed readers why surrogate models are needed for the design of mechanical systems. Then, the definition of a surrogate model and several basic related concepts were introduced, along with the nomenclature used in this book.

Chapter 2 reviewed five classic types of surrogate models, from traditional polynomial response surface models to neural networks, which are widely used in machine learning. Gaussian process models are also a representative type of surrogate models, due not only to their inherent characteristics in terms of function fitting, such as freedom from noise and the ability to provide prediction errors at unsampled points but also to their extensive usage in multi-fidelity surrogate modelling. The training process for a neural network was illustrated step by step, together with the derivation of the backpropagation algorithm for model tuning.

In Chap. 3, ensembles of surrogate models, which integrate the merits of various individual surrogate models, were studied as a special class of surrogate models. Unlike the classic surrogate models discussed in Chap. 2, there is no strict mathematical derivation for ensembles of surrogate models, and their predictions simply rely on the performance of each individual surrogate model. The key problem in the construction of ensembles of surrogate models is determining how to identify surrogate models with higher accuracy and assign higher weights to them. The weight of each individual surrogate model can be calculated based on either their

P. Jiang et al., *Surrogate Model-Based Engineering Design and Optimization*,
Springer Tracts in Mechanical Engineering,
https://doi.org/10.1007/978-981-15-0731-1_8

global performance or their local prediction accuracy at each sample point; these two approaches result in constant or pointwise weight coefficients, respectively, for each surrogate model throughout the design domain.

Chapter 4 presented a class of surrogate models that has been newly developed in recent years—multi-fidelity surrogate models. Unlike traditional surrogate models and ensembles of surrogate models, a multi-fidelity surrogate model considers multiple data sources of different fidelities. A multi-fidelity surrogate model fuses information from data of different fidelities by utilizing low-fidelity data to capture the general trend of the responses of the high-fidelity model while applying relatively few high-fidelity data to ensure the modelling accuracy. Multi-fidelity surrogate models take advantage of both the low simulation cost of low-fidelity models and the high accuracy of high-fidelity models; thus, they can achieve higher accuracy at a lower simulation cost. Three commonly used types of multi-fidelity surrogate models were studied, including scaling-function-based approaches, space mapping approaches and co-kriging approaches. A comparison of these three different forms of scaling functions was conducted based on numerical examples with different features, and the results showed that no form is universally better than the others. According to the 'no free lunch' theorem (Ho and Pepyne 2002), no surrogate model should be superior to all others for every application; thus, the selection of surrogate models should be conducted in a problem-dependent manner.

Once a surrogate model has been obtained, its prediction accuracy should be verified to give the designer some confidence that this surrogate model is well representative of the simulation model it replaces. Chapter 5 first presented the general model verification framework and classified the existing error metrics into those that rely on testing methods and those that rely on sampling methods, based on whether additional test points are needed in the model verification process. Then, several classic metrics in each category were introduced, and a review of the use of various error metrics for different application purposes was also provided. Finally, the performances of the error metrics for four classic surrogate models were studied to give readers an overall sense of which error metrics should be selected for specific types of surrogate models. Various influencing factors were considered in this comparison, including the number of samples, the noise level and the sampling method. Readers who wish to know which error metric among a set of alternatives is the best for a specific type of surrogate model can follow similar procedures to select the most appropriate metric for further design and optimization.

Another important issue related to surrogate models is the sampling method use. Chapter 6 presented several commonly used one-shot sampling methods and adaptive sampling methods. Given the same sample distribution, the sampling method has an important influence on the accuracy of the constructed surrogate model, especially when the samples are sparse. Adaptive sampling methods select sequential sample points based on input space information and/or output space error information; thus, the samples can be allocated more reasonably, and the surrogate model can be effectively made to reflect more properties of the system at the same simulation cost. Moreover, an adaptive sampling criterion can be based on either exploration or exploitation of the design space or can strike a balance between them, depending on whether the purpose is optimization or modelling. It should be noted that sequential sampling methods for multi-fidelity modelling have rarely been studied by other researchers, and compared with those for single-fidelity surrogate models, the criteria for selecting sequential samples for multi-fidelity models are more complicated. The code levels and corresponding locations of the samples must be decided during the sampling process; thus, in addition to the input/output information of the model at each level of fidelity, the cost ratio and numbers of sequential samples for models of different fidelity also need to be considered.

Chapter 7 demonstrated several classic application modes of surrogate-model-based design and optimization. For deterministic optimization, three efficient global optimization (EGO) algorithms were studied, for both unconstrained and constrained optimization problems, and the procedures of these surrogate-based optimization (SBO) algorithms were illustrated in detail. In the context of robust optimization, the combined effect of the surrogate model prediction uncertainty and the design variable uncertainty was derived, and the merits of surrogate-model-based optimization were illustrated by means of a practical design problem for a long cylindrical pressure vessel. Multi-objective evolution algorithms are another important application of surrogate models since the fitness evaluations generally require thousands of simulations and utilizing surrogate models in the optimization process can greatly reduce the simulation cost; sometimes, the cost reduction ratio may exceed 95%. Several novel surrogate-based multi-objective algorithms were introduced, and six numerical examples were presented to demonstrate their performance.

In summary, when applying surrogate-based design and optimization for the design of real mechanical systems, designers can generally apply the following procedures to build appropriate surrogate models and conduct the optimization process: the characteristics of the problem should be analysed first. Based on the dimensionality of the problem, the nonlinearity of the responses and other features, the most suitable optimization algorithm and type of surrogate model can be chosen. If the design budget is very limited, the adaptive sampling method can be used, and the space-filling criterion should be selected based on the optimization problem. Once the surrogate model has been obtained, its accuracy should be verified. For this purpose, the available error metrics should be compared, and the prediction uncertainty of the surrogate model should then be estimated using the most appropriate error metrics. Finally, the surrogate model can be integrated into the optimization framework, and the optimization algorithm can be run. At this point, a reasonably reliable optimal solution should be obtained, and the designers can compare the designed model against the available experimental results for further validation.

Reference

Ho Y-C, Pepyne DL (2002) Simple explanation of the no-free-lunch theorem and its implications. J Optim Theory Appl 115:549–570

The manufacturer's authorised representative in the EU is Springer Nature Customer Service Centre GmbH, Europaplatz 3, 69115 Heidelberg, Germany. If you have any concerns regarding our products, please contact ProductSafety@springernature.com

Printed and bound by CPI Group (UK) Ltd, Croydon, CR0 4YY
25/09/2026
02222900-0007